T0345014

Empowering Women in STEM

Throughout the globe, STEM careers exist. However, in some countries, particular STEM careers have been male-dominated while in other countries no gender biases exist as it relates to STEM careers. One common trend which occurs throughout the world is that women who are working in these STEM related fields typically leave after about five years. Contrarily, it has also been uncovered that more women stay in these fields when we all work together.

Empowering Women in STEM: Working Together to Inspire the Future provides a platform to share the stories of those who have been in STEM careers but have pivoted to other areas by utilizing the STEM skills they learned. It bridges the gap between those who are thinking about entering or leaving STEM careers, along with those who want to encourage others into STEM careers. This book showcases how everyone's journey is different, some may have unexpected twists and turns while others appear to conform to the "normal" rules outlined by society. By offering a front-row seat on a journey that takes many different paths, this book provides advice that can lead to a STEM career with or without having a STEM background. The different roads taken are highlighted to show how everyone's path is unique and how that is okay.

With the upcoming generation constantly looking for ways to "fit in" or be able to identify with role models to help them chart their way forward, this book ensures that they have not just one, but a variety of role models and success stories to relate to. It also offers some key advice which can be applied to any field they choose. In addition to having women and men from across the globe share their stories about various fields, this book also is written for professionals who may be considering a switch of career or deciding to leave STEM, and for university students who are trying to figure out their career choices and paths to take to gain more insight into possible new career goals in STEM.

Empowering Women in STEM
Series Editor: Sanya Mathura

This new book series aims to encourage more women into STEM-related fields and will provide a source of inspiration for those wanting to join or remain in these fields. In the last decade, it has been extremely difficult to get documented information on experiences women have faced in male-dominated fields. Typically, this information has been frowned upon by colleagues and women have been subjected to believing that they are "not strong enough" for these types of industries. As a result, there is a higher number of women leaving STEM fields after the first few years.

This series will serve to encourage more women to embrace their unique abilities and find comfort in knowing that other women have endured similar situations and not only survived but thrived. Too often, the media selectively chooses the highlights of these careers but not the struggles behind them. Essentially, these struggles are what get women prepared to achieve amazing things. However, if no one is enlightened about the highs and the lows of a successful STEM career journey, then other women in similar situations may feel that giving up is the easiest solution.

The series will also serve as an avenue for more male involvement in becoming allies and informing others about ways to bridge these gaps. Quite often, male colleagues are not aware of the struggles women face or that there are ways to help in these situations. Imposter syndrome is very prevalent in this area for men, but it is only when both men and women work together that the scales can be balanced. This series is targeted towards professionals new to the field and just starting a career in STEM as well as professionals who are already in these fields but may be considering leaving the industry. This series will also provide insight to companies trying to understand the role of diversity and inclusion from an employee's perspective and can help in shaping a company's culture. Additionally, it will provide those who are not within the STEM fields with much-needed insight into ways they can support someone within these fields.

If you are interested in writing or editing a book for the series or would like more information, please contact Cindy Carelli, cindy.carelli@taylorandfrancis.com.

Empowering Women in STEM
Working Together to Inspire the Future
Edited by Sanya Mathura

Empowering Women
in STEM
Working Together to Inspire the Future

Edited by
Sanya Mathura

CRC Press
Taylor & Francis Group
Boca Raton London New York

CRC Press is an imprint of the
Taylor & Francis Group, an **informa** business

Designed cover image: Sanya Mathura

First edition published 2024
by CRC Press
2385 NW Executive Center Drive, Suite 320, Boca Raton FL 33431

and by CRC Press
4 Park Square, Milton Park, Abingdon, Oxon, OX14 4RN

CRC Press is an imprint of Taylor & Francis Group, LLC

© 2024 selection and editorial matter, Sanya Mathura; individual chapters, the contributors

Reasonable efforts have been made to publish reliable data and information, but the author and publisher cannot assume responsibility for the validity of all materials or the consequences of their use. The authors and publishers have attempted to trace the copyright holders of all material reproduced in this publication and apologize to copyright holders if permission to publish in this form has not been obtained. If any copyright material has not been acknowledged please write and let us know so we may rectify in any future reprint.

Except as permitted under U.S. Copyright Law, no part of this book may be reprinted, reproduced, transmitted, or utilized in any form by any electronic, mechanical, or other means, now known or hereafter invented, including photocopying, microfilming, and recording, or in any information storage or retrieval system, without written permission from the publishers.

For permission to photocopy or use material electronically from this work, access www.copyright.com or contact the Copyright Clearance Center, Inc. (CCC), 222 Rosewood Drive, Danvers, MA 01923, 978-750-8400. For works that are not available on CCC please contact mpkbookspermissions@tandf.co.uk

Trademark notice: Product or corporate names may be trademarks or registered trademarks and are used only for identification and explanation without intent to infringe.

Library of Congress Cataloging-in-Publication Data
Names: Mathura, Sanya, editor.
Title: Empowering women in STEM : working together to inspire the future / edited by Sanya Mathura.
Description: First edition. I Boca Raton, FL : CRC Press, 2024. I Series: Empowering women in STEM I Includes bibliographical references and index.
Identifiers: LCCN 2023041694 (print) I LCCN 2023041695 (ebook) I ISBN 9781032679495 (hbk) I ISBN 9781032678948 (pbk) I ISBN 9781032679518 (ebk)
Subjects: LCSH: Women in engineering. I Women in science. I Women in mathematics. I Women--Professional relationships. I Engineering--Vocational guidance. I Science--Vocational guidance. I Mathematics--Vocational guidance.
Classification: LCC TA157.5 .E46 2024 (print) I LCC TA157.5 (ebook) I DDC 620.0082--dc23/eng/20231208
LC record available at https://lccn.loc.gov/2023041694
LC ebook record available at https://lccn.loc.gov/2023041695

ISBN: 978-1-032-67949-5 (hbk)
ISBN: 978-1-032-67894-8 (pbk)
ISBN: 978-1-032-67951-8 (ebk)

DOI: 10.1201/9781032679518

Typeset in Times
by SPi Technologies India Pvt Ltd (Straive)

Contents

SECTION I   *Inspiring the Future*

SECTION II   *Non-Traditional Paths*

SECTION III   *Leading the Way*

About the Editor

Sanya Mathura is the Founder of Strategic Reliability Solutions Ltd based in Trinidad & Tobago.

She works with global affiliates in the areas of Reliability and Asset Management to bring these specialty niches to her clients. She has a BSc in electrical and computer engineering and an MSc in engineering asset management from the University of the West Indies. Sanya has worked in the machinery lubrication industry for the past decade and assists various industries with lubrication-related issues.

Sanya is the first person (and first female) in her country and the Caribbean to attain an ICML MLE (International Council for Machinery Lubrication Machinery Engineer) certification as well as the first female in the world to achieve the ICML VPR and VIM badges. She is the author and co-author of five books; *Lubrication Degradation Mechanisms – A Complete Guide, Lubrication Degradation – Getting into the Root Causes, Machinery Lubrication Technician (MLT) I & II Certification Exam Guide* and *Preventing Turbomachinery "Cholesterol" – The Story of Varnish.* She also edited the book *Empowering Women in STEM – Personal Stories and Career Journeys from Around the World* and is the series editor for this book series. When not writing or managing the business, you can find her supporting projects to advocate for women in STEM.

Sanya's passion for excellence, coupled with her expertise in the field of engineering and reliability, has made her a respected and highly sought-after professional in the industry. Her dedication to providing exceptional service to clients and her commitment to staying up to date with the latest industry trends has earned her the respect of her peers and the admiration of her clients.

Contributors

Jenny Ambler
UPTIME Consultant Ltd
United Kingdom

Erika Anderson*
Georgia Pacific LLC
USA

Sarah Marie Bilger
IPS – Integrated Project Services
USA

Shelli Brunswick
Space Foundation
USA

Marcella Ceva
WE Ventures – Microsoft
Brazil

Genevieve Cheung
Atticus Consulting
Canada/Barbados

Ludmilla Derr
Elite Experts Conferences Ltd
Switzerland

Claudia Gomez-Villeneuve
MacEwan University
Canada

Angelica González
Hexpol Compounding
Mexico/USA

Kenya L. Goodson
Hometown Action
USA

Stephanie Hajducek
This One's for the Gals
USA

Corey Marie Hall
STEM Education Works
USA

Viktoria Ilger
Creators Expedition - an AVL Initiative
Austria

Helen Sara Johnson
Manufacturing Technology Centre
Loughborough University
United Kingdom

Alex Knight
STEMAZING
United Kingdom

Joel Leonard
MakesboroUSA
USA

Susan Lubell
Steppe Consulting Inc
Canada

Eman Martin-Vignerte
Bosch UK & Ireland
United Kingdom

Katie Mehnert
ALLY Energy
USA

Stuart Naismith
North Lanarkshire Council
United Kingdom

Lennis Perez
Just Lennis, LLC
USA

Shane Woods
Girlstart
USA

Emily Soloby
Juno Jones®
USA

*Sunrise (12-5-89) and Sunset (9-20-23)

Erika Anderson

Erika was introverted, contemplative, and studious. Her preschool teachers repeatedly bragged that at 3 years old, they felt she was brighter than their other students, so much so that they predicted she would grow up to be a doctor or lawyer. As a child, Erika didn't want to go outside and play like the other kids but preferred to stay indoors reading and when asked about it, she responded, "Unlike my peers, I enjoy learning." From elementary school years onward, Erika was consistently rated an exceptional student and for the remainder of her primary, middle, and high school education was placed in "high achievement" classes earning straight As in all classes. It wasn't until she attended Spelman College from where she graduated with honors undertaking dual degree programs in STEM, a BS in Science, and a BS in Math from Georgia Tech, that she truly blossomed into the passionate, outgoing, free-thinking, and well-rounded woman she embodied as an adult. While working as a Mechanical Engineer with ExxonMobil in Texas, she earned a master's degree in Data Analytics. Erika is missed terribly, but we hope that her legacy, her passion for learning, and the accomplishments that she leaves behind will inspire others to pursue similar endeavors.

Jay, Liz, and Pamela Dommond

Foreword

Katie Mehnert
ALLY Energy, USA

What would you do if you weren't afraid?

It's a question Sheryl Sandberg, friend, former COO of Meta, and "Steminist", asked me 10 years ago when I read her book, *Lean In*. I didn't know how to answer the question. It made me take a serious pause about how I wanted to make a difference in my career and in the future.

As we turn the pages of this remarkable book, let's pause for a moment to acknowledge the strength and resilience of the countless women and men who have courageously ventured into the fields of science, technology, engineering, and mathematics (STEM). Their journey, often marked by challenge and adversity, is a testament to their enduring spirit and drive.

They didn't let fear stop them. They answered the question and the call to act!

These women have dared to penetrate the traditionally male-dominated disciplines by challenging the status quo and inspiring countless embryonic minds to follow in their footsteps. They are not just scientists, engineers, mathematicians, or technologists; they are visionaries, trailblazers, and game-changers. The men featured in this book are important allies. An ally is a force for good. They have supported women along their journey and aided in encouraging more men and women to become part of this remarkable adventure.

In the wake of the climate crisis, these women and men have risen to the fore. They are utilizing their skill sets and knowledge in STEM fields to address the multifaceted issue of climate change. Through the use of mathematics to model climate patterns, engineering to design renewable energy systems, science to understand the impact of human activity on the environment, and technology to develop sustainable solutions that reduce our carbon footprint, they are all paving the way forward for future generations.

They are at the heart of a green revolution where sustainable practices meet technological advancement. They are leading the charge on energy conservation, ushering in a new era of renewable power sources, and driving the immeasurable potential of clean technologies. From harnessing the potential of wind, solar, and hydropower to exploring nuclear and bioenergy, women in STEM fields are at the forefront of changing how we produce, distribute, and consume energy.

While they navigate this complex landscape, they also act as mentors, inspiring young girls, and boys, to explore the thrilling world of STEM, to question, innovate, and dream without limitations. They are the embodiment of the idea that anyone, regardless of gender, can contribute to solving the problems that threaten our world today.

But their work is far from over. We need more like them. We need you!

Yes, I said, *"We need you!"*

As we venture into the uncharted waters of the future, the demand for STEM expertise will only grow. Our planet's survival depends on sustainable practices, innovative technologies, and a global commitment to a green economy. This book provides a platform to recognize and celebrate the incredible work done by women in STEM fields.

In reading their stories, their struggles, and their victories, may we all find inspiration. Their dedication to their fields, their commitment to sustainability, and their tireless work to combat climate change are nothing short of empowering. In their hands, the future of our planet is not just a problem to be solved, but a promise to be fulfilled.

And so, I challenge you to rise to the occasion. Lean into this great and rewarding opportunity. In your hands, you hold a book that will empower you to effect change and shape the future of our planet. Be curious. Be daring. Be relentless in your pursuit of knowledge.

The question is, *are you ready to put aside the fear and answer the call?*

ABOUT THE AUTHOR

The modern architect of the energy workforce, **Katie Mehnert** is the Founder and CEO of ALLY Energy, the community accelerating connections, jobs, and skills to drive an equitable energy transition. She was appointed Ambassador to the United States Department of Energy in 2020 and has testified before Congress on the clean energy workforce of the future. She was most recently appointed to the National Petroleum Council. She's also an Energy Institute Fellow and an advisor to Clean Energy for America. Katie is a speaker, author, and trusted source in the energy industry. She has been published in *Scientific American*, *Forbes*, *The Hill*, *CNBC*, and *CNN*. Her first book, *Grow with the Flow*, was published in 2020. She most recently co-authored *Everyday Superheroes: Women in Energy*, a children's book focused on energy careers. Katie also has appeared in *Hot Money*, a documentary produced by Academy Award-winning actor, Jeff Bridges and Retired NATO General Wesley Clark on the financial complexities of climate change and finance, and *Dirty Nasty People*, a film on the future of the energy workforce. Katie is a four-time World Major marathoner having completed London, Chicago, New York, and Berlin. Her husband is a legal executive with Baker Hughes. They live with their 12-year-old daughter, Ally, in Houston.

Preface – Women in STEM

This book is a compilation of stories told by both men and women who have worked in the STEM fields of science, technology, engineering and mathematics. Some of the authors do not have a STEM background but have made significant strides in the STEM world, while others have used their STEM background to make an impact in other areas such as entrepreneurship and even finance. In most of these cases, they all moved forward by working together.

There are three main sections in this book: Inspiring the Future, Non-Traditional Paths, and Leading the Way. Each section takes the reader on a journey of ways in which the authors have patterned their lives to inspire future generations, from deep space exploration to daring to be different. Some of the authors also give insight into ways they have gotten into non-traditional paths, including being the first wellness engineer or creating an avenue for children to enjoy science while learning the fundamentals. Other authors describe how they have led the way by creating new paths towards education or in fields which may not have existed ten years ago.

By making their impact in STEM careers or paving the pathway forward for others, these authors have all shared a piece of their soul and the work that they are doing to help inspire future generations by working together. It was very important to ensure that this book showcased chapters written by men as well. As such, the audience of the book is not just limited to women. The only way to move forward and inspire the future generation is by all of us working together and building new paths for others to follow.

Overview

INSPIRING THE FUTURE

In the STEM world, women face a couple of challenges, but they are not just limited to this earth, there are also challenges that women encounter in space. Our first author, **Shelli Brunswick**, describes some career-defining lessons for women in space based on her experience and that of several other women. A female professor, **Alexandra Knight** from the United Kingdom, shares her journey on being the change that you want to see in the world and how this made her develop STEMAZING, especially during the pandemic. From Kazakhstan to Germany to Switzerland, we get a first hand account of daring to be different and being true to yourself, as **Ludmilla Derr** uncovers her journey in tech and what it means to develop healthy working habits.

Female engineers often face some difficulties; this includes Arab female engineers, and one of our authors, **Eman Martin-Vignerte**, recounts ways to survive and thrive in this industry as she has paved the way forward for others. Another author, **Stephanie Hajducek**, describes her journey towards fueling the future of women in the industry through her organization, 'This One's for the Gals', where high school students are given the opportunity to participate in STEM activities. Hailing from Brazil, author **Marcella Ceva** focuses on building the next generation of STEM heroes through her work at Microsoft to invest in more women in tech.

NON-TRADITIONAL PATHS

Sometimes life can take you on non-traditional paths. Our first author in this next section, **Stuart Naismith**, talks about not having a background in STEM but going on to teach science to students while making it fun. He wanted to make more of an impact in this world and decided to equip future generations with the tools that they would need. Even if someone comes from a banking background, this does not mean that they cannot enter a STEM field or the reliability sector to be more exact. **Jenny Ambler** shares her journey on moving from banking to becoming the first female technical operator at the Doritos factory in Turkey.

Making the transition from being a bookworm to microbiology to coding to eventually becoming a librarian and a curriculum innovator, another author, **Corey Marie Hall**, shares her very intriguing story of unexpected paths and endless possibilities. Our careers can often feel like a moving sidewalk, and **Susan Lubell** provides some advice on ways this can be approached as well as the importance of mentorship and networking. Being able to explore innovation, STEM and entrepreneurship is a journey of learning and growth as explained by our next author, **Viktoria Ilger** from Austria. She recounts her entry into the STEM world and working with start-ups.

All of our authors have been breaking glass ceilings but one in particular uses safety boots to do so! With a background in law and a passion for helping women, **Emily Soloby** gives us some insight into how she began manufacturing safety boots designed for women. Quite often, we get thrown into a headwind and can't seem to find our way out or forward. *Erika Anderson* explains how she changed some head-winds into tailwinds to help guide her career.

A wellness engineer was not a job that existed ten years ago. Even today, it is very new territory. *Lennis Perez* gives us the behind-the-scenes footage of how she developed this concept and is utilizing it today. There is an intersection of science, education and policy, but very few are able to find it and leverage its benefits to others. Our author *Kenya L. Goodson* provides us with her nonlinear journey into this specialized area.

LEADING THE WAY

Women often lead the way, and *Sarah Marie Bilger* talks about redefining work-life balance in STEM and bringing parenthood into the equation. She even talks about some strategies which can be utilized. Often, there is a gender occupation dilemma for females in engineering. *Helen Sara Johnson* explores this concept of fitting in without fitting in and the relationship between identity, authenticity and work outcomes.

Adults aren't the only ones who face challenges in STEM; kids do as well! *Shane Woods* provides some info on ways to make STEM more accessible to girls and the conversations we should be having with caretakers. Making the move from Colombia to Canada to study engineering and maneuver the path of project management while raising a family is laid out for the readers by *Claudia Gomez-Villaneuve*. She talks about ways to continue earning PDUs on the way to becoming a registered professional engineer.

Our next author, *Angelica González*, talks about the benefits of growing up in a traditional Mexican family and how it has shaped her into the person she has become today. We are all aware of the underrepresentation of women in STEM, but *Genevieve Cheung* takes this a step further to explore gender diversity, the Equality Agenda versus the Equity agenda and ways to shift the narrative. Our final chapter focuses on the importance of empowering women in STEM from a male perspective, as *Joel Leonard* gives insight into ways he has helped women in China, India, Nepal, Moldova and even Ukraine to develop their STEM careers.

All of these stories are very inspirational and help us to better understand how we can all work together to inspire the future.

Acknowledgements

It is my great honor to work with such an elite group of authors who have all contributed to by sharing their stories in this book. These leaders have graciously allowed us to learn from their journey as they broke barriers, created new professions, and led the way. A huge and sincere thank you goes out to everyone who has taken the time to share your journey as we all work together to inspire the future.

Section I

Inspiring the Future

1 Space for Women
Defining Career Lessons

Shelli Brunswick
Space Foundation, USA

1.1 OPENING DOORS AND ENSURING WOMEN WALK THROUGH THEM

The space race began as a logical progression of the eternal quest for advantage over our international rivals that comes from holding the high ground. It has expanded into wondrous achievements of exploration and invention that have transformed our world and our lives. Private enterprise has seized the opportunity and is fueling growth, limited only by the same weakness that threatens the future of innovation: workforce development.

In 2022, "woman" was the word of the year, according to Dictionary.com. We've seen it attached to many efforts that place specific emphasis on bringing women into the fold to bolster the workforce. For example, "Space and Women" was the theme for the 2021 World Space Week; "Women and Technology" is the focus of the WomenTech Network; and "Women and Entrepreneurship" was the subject of a global study conducted in 2022 by the World Business Angels Investment Forum (WBAF) on the state of female entrepreneurs (Elsesser, 2022).

While there most definitely are not yet enough women in STEM, these examples show a willingness not only to open the door, but to ensure women walk through, because women are essential to creating the workforce needed to drive innovation forward. I can tell you firsthand that there is space for women, in the space industry in particular.

To this end, I want to use this chapter to highlight lessons learned throughout my decades-long career in the space industry. I'm going to walk you through what these lessons looked like, through my own experiences as well as those of other amazing women I've been fortunate enough to know. Then, I will discuss the role of failure and the importance of mentorship in achieving your goals and fulfilling your purpose.

Now let's dig in.

1.2 LIFE'S LESSONS

You likely picked up this book to find new ways to pursue your dreams. Some of you may just be starting your career journey; the future lies in front of you, and you're looking for insight on what to do next. What career paths and opportunities might be suitable for you to follow? And who might be able to assist you? Others may be

DOI: 10.1201/9781032679518-2

looking to make a change, find a new pathway, and start a new adventure as you look to turn the page. Everyone's story is different.

I'll tell you, my personal career journey (so far) can best be described by three chapters. The first focused on when I enlisted in the United States Air Force (USAF) out of high school. The second is when I became an officer and started my career as a space professional. And the third chapter is devoted to my retirement from the USAF and becoming the Chief Operating Officer at Space Foundation. As a side note, I concurrently serve as the G100 Global Chair for Space Technology and Aviation as well as a WomenTech Network Ambassador—I mention this because it goes to show that career chapters can be rich and extend beyond traditional boundaries of a story.

Many of the lessons I will share with you here, though, come from the first and second chapters of my career, because I am still writing the third! Now, let's start with chapter one.

I enlisted in the USAF right out of high school. I took this route for a few strategic reasons: one, I wanted to see the world; two, I didn't have money to go to college; and three, I didn't know what I wanted to major in even if I had the money for school. I had no real idea what I wanted to do with my life beyond experiencing new things and maybe learning some skills. Pretty common themes for an 18-year-old, even today.

The USAF was my best option, and I am grateful for that. It offered me an opportunity to see and experience things I could have only imagined back home. While in the USAF, I had the privilege to be stationed in Turkey, Germany, and the United States Air Force Academy in Colorado Springs. I treasured discovering other cultures, customs, and people. In addition to the intangibles that come from living outside of my norm and experiencing life abroad in two very different countries, I also gained a true skill set that would benefit me throughout my career chapters to come.

As an enlisted airman, I worked as a personnel or human relations specialist. This taught me all about customer service, promotions, training, and benefits even if my position didn't translate directly to the typical HR job in the private sector. I also qualified for the GI Bill, which allowed me to finish my college education without the crippling debt of student loans. But some interesting things happened along the way as I matured and learned more about who I was and what I hoped to achieve.

1.3   TAKE ADVANTAGE OF EVERY OPPORTUNITY

I don't like to see an opportunity pass me by—even if it makes life more challenging at the moment. As an enlisted airman, I had the chance to utilize tuition assistance and complete college courses in the evenings and on weekends, and I knew I had to take advantage. This was an investment in my future! As uncomfortable as it might have been over the short term (I missed a lot of parties and a lot of sleep), what could be more important?

Furthermore, many of my Air Force training programs offered college credit. So although my intention had been to join the U.S. Air Force to earn college money, I ended up completing both my bachelor's and master's degrees as I served my country. In fact, I still qualify for the GI Bill and look forward to enjoying its benefits again at some point.

This is just one example of how taking advantage of every opportunity has helped me. Did I know in my late teens/early 20s that taking some classes would translate to multiple degrees and becoming a lifelong learner? Absolutely not. But I did think education could be a valuable tool for my future, and I shouldn't turn down the chance to have someone else pay for it.

Many other global leaders in the space industry also believe in taking advantage of every opportunity. Perhaps it is hardwired in us somewhere—the desire to explore all of our options without regard for traditional constraints. By taking advantage of even small cracks at openings, we've not only changed our own lives but the lives of those around us as well. Dr. Iroka Chidinma Joy is an example of someone who did exactly this.

1.3.1 CHANGING OUR LIVES AND THOSE OF OTHERS

Encourage people to grab any opportunity when it comes.

—**Dr. Iroka Chidinma Joy**

Dr. Iroka Chidinma Joy is a leading woman in the African space industry. A scientist, mentor, and role model, she holds the position of Assistant Director of Engineering at the National Space Research and Development Agency (NASRDA), which is the national space agency of Nigeria. She earned her PhD from Ladoke Akintola University of Technology (LAUTECH) in Ogbomosho, Oyo State, Nigeria, in the Electronic and Electrical Engineering Department.

Iroka believes her journey into the space industry was divinely appointed as per our Zoom interview, 2022a. She possessed a talent for technology, drawing, and crafts; however, she never dreamt that she would have the opportunity to pursue a space career. She is of the faith that you don't need to know exactly where you are headed; just be patient and trust the process.

Iroka's journey into space began when she first attended a polytechnic institute to study electrical and electronics engineering, following which she pursued a post-graduate degree in electrical engineering. Her successful completion of a written examination for a NASRDA job launched her into the space sector.

Her role at NASRDA has been intriguing because it has given her room to conduct research and apply knowledge in her field (Interview and AfricanNews, 2020). Iroka's experience was unique because she was the only woman on her team of Nigerian engineers and scientists. She was included in every step of the process of building and testing spacecraft—something very few people ever have the opportunity to do.

When we spoke about our career trajectories, Iroka mentioned that she didn't have a mentor when she started out. She believes that if she'd had one, she would be even further along in her impressive career, breaking even more boundaries. But what worked in her favor was a strong belief in herself and her own abilities. She thought she could do it, and she did! She learned independence and self-reliance; she was willing to take on tasks and projects that she didn't think she was up to, even when they weren't paid work. She saw volunteering as an opportunity to learn. When she didn't know how to do something, she would research it and figure it out.

Iroka has never said no to opportunity. She bravely took on challenges, and as a result, her confidence grew. She is now a mentor and a role model to the next generation of space leaders in Africa, and she's well aware that many people are looking up to her. "Opportunity did not come early in my life; that is why I advocate for young girls and boys to come into the space industry," she says. She wants such decisions to be easier for them; she wants to provide representation and hope that space is an industry open to anyone.

Iroka is passionate about having an impact on others through her efforts, and she has her own suggestions for creating more opportunities for the next generation of women in space. First, she recommends that various stakeholders, such as the Nigerian Space Agency, the United Nations Office of Outer Space Affairs (UNOOSA), the African Union Commission (AUC), and the Islamic World Educational, Scientific, and Cultural Organization (ICESCO), host an "African Space for Women" Expert Meeting. This Expert Meeting would focus on creating pathways for women and girls in the space ecosystem, enhancing partnerships, fostering workforce development and capacity-building, and promoting efforts to encourage women and girls to get involved in STEM and space-inspired education (Space in Africa, 2020).

Her second suggestion is to reform elementary and middle school education. Schools need to build awareness about space opportunities, yet most science teachers lack sufficient background in space education. Therefore, teachers need training that includes hands-on learning and examples of how technology and space are used in the real world. This would increase awareness about the space ecosystem and how space benefits our daily lives. Most students simply don't know about aspects of the space ecosystem that pertain to space exploration, technology transfer, and space-to-earth applications. They have not been exposed to the transformation taking place in the space industry and how it is creating new opportunities in Nigeria and across Africa. Success hinges on securing public-private funding from the government, space companies throughout the supply chain, and other stakeholders to improve the quality of access by girls and boys to STEM education as well as increase their visibility in the space community (Space in Africa, 2020; Zoom Interview, 2022a).

What I hope you glean from my experiences in the USAF and Iroka's unexpected path to the space industry is that by taking advantage of every opportunity, we become more proactive in our own lives. By seizing opportunities, we take control of our destiny, often in very unexpected and rewarding ways.

1.4 TRY ANYWAY, EVEN WHEN THE ODDS ARE AGAINST YOU, DON'T GIVE UP

I will sink a little deeper into my time in the USAF for my next lesson. As an enlisted airman, I realized I wanted to be an officer. I met the requirements after I got my bachelor's degree. However, the Air Force was looking for STEM professionals in science, technology, engineering, and mathematics, and I was none of these things. My bachelor's degree was in business administration.

So, my odds of being selected were only about 12%. I didn't let that stop me, though. I finished my application and got all my transcripts and letters of recommendation together. And I waited months to hear the results, only to learn that I was not selected.

Although I was disappointed, the Air Force had a process at that time for an unsuccessful application to be looked at by the next officer selection review board. Applicants had the option to update their applications. I chose to do so because I believe one should always present oneself in the best possible way. As a result, I requested that the incredibly supportive people who had written my letters of recommendation update their letters, which they did. I also made sure that all the information was up to date and rewrote my essay before resubmitting my application. After that, I thought about my options as an enlisted airman and came up with Plans B, C, and D for my career. I started to consider alternatives and what my next steps would be to achieve those goals if I were denied again. Fortunately, I didn't have to turn to those options, but it gave me a sense of calm and control to know that I had backup plans that I could live with.

When I received the news from the second selection board that I had been chosen to become an officer in the USAF, I was elated. During this process, I learned to "try anyway"—even when the odds were stacked against me. Let this be a lesson to you, too! If you really want something, don't give up—keep going. Build your persistence, tenacity, and resilience, because you may have to try more than once to reach your goal. Instead of fearing failure, use it to learn, grow, and develop. You will be better for it, trust me.

Countless other leaders have experienced setbacks and delays, but they also didn't give up. They were likely more successful in the end because of all the work it took to reach their goal, all of the learning and understanding that came throughout the process.

1.4.1 Success Doesn't Just Depend on the Goal but the Process

> I believe that dreams can come true. It doesn't matter how crazy that dream is or how long it takes to achieve it; with perseverance, if you believe in it, you will achieve it.
> —**Priscilla Nowjewski**

Priscilla Nowajewski is a data analyst who works at the Atacama Large Millimeter/ Submillimeter Array (ALMA). Located in northern Chile, ALMA is the most complex astronomical observatory ever built on Earth. It's only fitting that Priscilla is part of the dynamic team that works at ALMA, because she is also a very dynamic person, seeing herself as a scientist, teacher, entrepreneur, writer, watercolorist, and science communicator.

As per our Zoom interview (2022b), Priscilla notes that she had role models early on, like her female high school science teachers who helped her see herself as a scientist. She also saw role models on television shows such as *The X-Files*, where FBI agents Mulder and Scully were solving mysteries. Priscilla notes that she wanted to be like Dana Scully because she was confident and brilliant. These were traits that Priscilla acknowledges she didn't possess at the time, but she dreamed of having them someday. Her journey to find her true passion and destiny took some time.

In undergraduate school, Priscilla studied physics. But, this wasn't her first choice. She really wanted to study astronomy; she didn't get accepted into that program, however. Priscilla didn't let that stop her. She collaborated with a friend, and together they started knocking on doors and asking if astronomers would let them work for them. By her second year in undergraduate studies, she was able to study the extinction of light in star-forming zones with a very well-known Chilean radio astronomer. With determination, grit, and drive, she found opportunities where she could meet astronomers, learn from them, and develop her understanding and skills.

After completing her bachelor's degree, Priscilla started a master's and PhD program in physics because she was offered an opportunity to study at the Max Planck Solar System in Germany, where she had been working with the Comet group. They offered her the opportunity to do her PhD with them. Unfortunately, that didn't happen because of some challenges with a professor who wanted to have her expelled without giving a reason. She decided on her own to resign from this master's program and give up the chance to get her PhD. But, this change in career path allowed her to find her passion in planetary atmospheres. With this newfound sense of purpose, she entered a PhD program in fluid dynamics, where she specialized in atmospheric dynamics and planetary atmospheres.

There had been obstacles along the way, beyond the professor threatening expulsion. Physics had never been easy for her. It took her more time to finish her undergraduate assignments than other students. She realized that she didn't have all of the math skills she needed to do well in her college-level classes. She approached counselors and faculty at her university for help. They found out what courses and skills she didn't have and made a program to help her learn them. This allowed her to successfully finish her classes and get a physics degree.

During this time, she doubted her ability to pay attention and do her homework. She would say to herself things like, "I'm not stupid," "I know I can do it, but what's the problem?" and "Why can't I concentrate or perform in my studies?" That's when she decided to go to the doctor to see if there was another reason for her difficulties. She was diagnosed with ADHD, and instead of thinking it was a limitation, it helped her understand what was happening and, more importantly, what truly interested her. She finally understood why studying things that didn't pique her curiosity was so hard.

Hearkening back to *The X-Files* and Agent Dana Scully, Priscilla recognized that she was interested in UFO topics and had a special interest in looking for life on other planets. It allowed her to think about the possibility that we are not alone in the universe, and our solar system really caught her attention.

She remembered seeing Saturn through a telescope for the first time at an astronomy summer school at the University of Chile, where she later got her degrees. She could not believe what she was seeing—a planet with its rings so well defined that you could even see the gap between the rings. That memory remained in her mind, and years later she had the opportunity to participate in a Capacity Building Workshop on Planetary Science (COSPAR) in Uruguay with two other Chilean students. There, she had the opportunity to work with data from the surface of Mars and discovered dust devils, which really fascinated her. With her fellow Chilean cohorts, she studied the surface of Mars using stereoscopic images, which required the use of red and green glasses to see the 3D effect. As three Chileans, they proposed the idea of creating a Planetary Sciences Group in Chile as their final project. It was 2007, and in astronomy they talked only about stars, galaxies, and other objects; there were still no experts on planets in Chile.

The COSPAR workshop created ties and networks with other experts in planetary sciences, allowing Priscilla to collaborate with the University of the Republic of Uruguay Asteroids Group. At that time, they proposed that Pluto was not a planet, drawing a lot of global attention. It also gave Priscilla the chance to work with the German Comets Group. Through that partnership, she took part in NASA's international EPOXY campaign and studied comet Hartley 2 from Chile using the Gemini South Telescope.

Priscilla's planetary search and self-discovery of her true passion were again thanks to the planet Saturn. Before resigning from the physics master's degree, she took a course in planetary sciences taught by the first expert on planets in Chile. The course had a section on atmospheres, and she was shocked when she saw that Saturn had a hexagon at its North Pole. She couldn't believe it. It was a kind of permanent storm that had retained its shape for years since the first images of Saturn: the hexagon. She found her true passion, planetary atmospheres, because she needed to know why and how that hexagon was formed.

In 2011, she began her PhD in fluid dynamics. Even though she was doing much better at the university, she knew it would be difficult to stay focused on this topic because there were only a few people in the world who were dedicated to studying the atmospheres of other planets. It would be a significant challenge to be a pioneer in this field. But her studies got easier and easier, and when she finished her program, she knew that this was her true calling.

Priscilla is very grateful for what she achieved in her PhD, and her work has mattered. In fact, NASA invited her to present her results at a planetary sciences workshop at the Ames Research Center. She also had the opportunity to go to a summer school on paleoclimate in Italy, which allowed her to develop as a programmer. She even did a postdoc before finishing her PhD, due to her knowledge of astronomy and climatology—all things she never thought she would have the opportunity to do.

But one thing became clear: despite her expertise, planetary sciences still did not exist in Chile. She was overqualified for any job in her country, and it was even difficult to get postdocs abroad. She experienced a lack of firsthand opportunities; she met people who wanted to take advantage of her knowledge, and she made some bad decisions.

In a moment of great sadness, Priscilla realized how difficult it was to try to move forward in this field and was about to give up. But thanks to the support of her family and her partner, she found the resilience to move forward. Her partner had always been the one who encouraged her to continue; even when she had doubts, his faith in her never wavered. Instead of looking for opportunities, she started creating them. Priscilla founded The Mars Society Chile. Finally, she realized that she is the catalyst for her country's advancement in planetary sciences, and she wants to invite others in so that no one else experiences the isolation she had.

So now she's on a mission to create research opportunities centered on planetary atmospheres. In addition to her work as a data analyst, she continues to guide, mentor, and coach students who want to work with her on topics such as Mars, climate change, and exoplanets.

What Priscilla gained during this process was perseverance. You shouldn't give up, and you might have to try different things before you find your passion. She learned to try anyway, even when the odds were against her. There were many times during her schooling when she went to class and didn't have the complete answer to the homework problems. She figured out that we don't have to be perfect or have the complete answer to a problem. We just need to not give up and try anyway.

Priscilla also found humility, empathy, and leadership, as well as the need to be a champion for Earth. Through the Mars Society Chile, she created a community of like-minded people who wanted to expand science and climate awareness. She thinks it's important to talk about science and engage with the public about questions and findings; scientists have a duty to share their knowledge with the world.

With this mindset of building greater awareness, Priscilla collaborated with eight other writers from the Mars Society Chile to create *Vida: Su Origen, Vvolución y Búsqueda en el Espacio*, which translates to "Life: Its Origin, Evolution, and Search in Space," by Editorial Montacerdos. They wrote this book to help build understanding of the formation of life on Earth from the standpoints of astronomy, geology, climatology, and biology.

Because of Priscilla's passion to create greater awareness, she has embarked on a path to bring science closer to people. Through her Instagram account, she seeks to interact with the public so they can see that scientists are ordinary people who take curiosity to an extreme level; they are people who love to talk about all the things they discover. Priscilla is also starting a podcast called *Please, Can I Ask You Something?* This will eventually become a book, and she is collaborating with the ALMA community manager to share more stories about science, exoplanets, space exploration, and, above all else, the atmospheric conditions that must be in place for life to exist on other planets.

Priscilla has great faith in humanity and our ability to solve the problems posed by climate change. She believes that the development of space exploration and interplanetary travel will unlock innovation and technological advances that we'll be able to use to save our planet.

While I hope you've been inspired by Priscilla's story, as well as my own, the overarching lesson, "try anyway, even when the odds are against you," is important. It encourages persistence and resilience in the face of challenges or setbacks—which you will inevitably experience on your career journey. How you respond is up to you.

1.5   DON'T FEAR THE UNKNOWN

To help you move forward in moments of difficulty and uncertainty, I will tell you, don't fear the unknown. When I was selected to be an officer, I was assigned to specialize in space acquisition. I hadn't the faintest idea what this meant, and I certainly didn't want to do it. Instead, I wanted to be a personnel officer. It's what I was familiar with, and everyone I knew was in that career field.

I called several of my mentors and asked if they could help me, and they tried. They made several calls to attempt to reassign my career path from space acquisition to personnel. Finally, one day, someone from the Air Force Personnel Center called me and said, "Sergeant Brunswick, the Air Force needs you to be a space acquisition (program manager) officer." With that, I saluted sharply and said, "Thank you." And today, I have come to mean it from the bottom of my heart, because that started the next chapter of my life story. It led me to my position today, where I am able to share experiences with you. Because of that career pivot, I began my journey as a space leader, and I wouldn't have it any other way.

I had worked so hard to become an officer, and there were moments when I wasn't sure it would be worth it. But I owed it to myself to try, and I thought I should see what this whole space thing was about. Rather than run from it, I learned not to fear the unknown. I cannot advise you more strongly: be open to new opportunities and get used to moving outside your comfort zone. It is the best gift you can give yourself.

If you don't take my word for it, take Stela's.

1.5.1   Moving Outside Your Comfort Zone

Nothing is unchangeable. You can make a shift.

—**Stela Lupushor**

Stela Lupushor is the founder of Reframe, Work Inc., a consulting firm that advises clients on how to innovate and develop a workforce strategy that creates a resilient, inclusive, and accessible workplace by harnessing technology, human-centered design, and future thinking. Stela is also the founder of an amazing non-profit community group that helps women find new jobs and gives them the tools they need to excel in the workplace of the future.

In our zoom interview (2022c), Stela recounts her unique journey that started in the former Soviet Union, where she was born. She witnessed the dissolution of the Soviet Union and became a citizen of a new country called the Republic of Moldova. During this seismic change in both her personal and professional life, Stela was offered access to domains previously unthinkable. Many times, these new opportunities started with a question like, "Would you like to go?" or "Would you like to try?" She would accept this new, unknown adventure, and figure it out as she went along.

Initially, her new opportunities started in Moldova, where she had the chance to be part of the efforts to privatize state-owned enterprises and build the capital markets infrastructure for the new republic. This willingness to sample new things led her to move first to Ukraine and then to Russia to replicate her work in Moldova, namely, to commercialize and privatize government-owned businesses.

Some of the traits that enabled Stela to be successful were her curiosity and desire to develop her skills. She figured out process improvement, change management, service delivery, and strategy. Her influence continued to grow, and this opened up a new world to her, eventually resulting in an opportunity to move to the United States. "Life is about timing," she notes.

Stela was always willing to say yes to new opportunities because she didn't fit into a certain box, and change was the one constant in her life. She said,

Don't be afraid to make decisions today and worry that they will be the wrong ones or that they will put you on the wrong path for the rest of your life. I grew up in a country that doesn't exist. I spoke a language that doesn't exist. Nothing is unchangeable. You can make a shift. Look forward to what is coming.

You can adopt this mindset by being open-minded, asking questions, and being humble. "Challenges are always tough in the middle but beneficial at the end," Stela notes.

One of the reasons Stela is so excited about the global space ecosystem is because it is creating unprecedented opportunities. It brings about innovation and employment for many, challenges the human resources field to think creatively, generates economic impact, and influences the evolution of the digital workplace.

She currently teaches digital workplace design and design thinking as an adjunct professor at the New York University School of Professional Studies. In an example of this from one of her courses, Stela created a Mars colony case study for her students that capitalizes on the excitement about space innovation and exploration, allowing them to detach from earthly realities and think outside the box. The students must figure out how to reimagine the entire employment experience—attracting, engaging, developing, motivating, and rewarding employees—and innovate ways in which people can come to work on a Mars settlement. It's a very holistic case study. The hope is that some of the innovative thinking will translate into changes to today's work environment on Earth.

Her next challenge is to look at removing barriers that prevent people, especially women and minorities, from pursuing meaningful and gainful employment. These barriers can be personal (education, role models, risk tolerance, etc.), organizational (recruitment processes, rewards structures, credential requirements, etc.), infrastructural (access to childcare, transportation, broadband, etc.), and societal norms (biases and perceptions around accents, hair, color, body shape, and so on).

Another challenge she sees is ageism, especially for women. Societal norms make it more common for women than men to leave the workforce or decline professional opportunities to take care of children, ailing parents, and/or support a partner's career. Women who are re-entering the workforce often have to contend with outdated skills and stale business networks, which make it difficult for them to find employment or careers with long-term prospects. Biased job descriptions, i.e., "someone early in their career," and selection algorithms add to these challenges. The difficulties women experience in finding work at an older age impact their immediate income needs, which makes it challenging to retire. Less time at work equates to less personal savings and smaller social security payments. If we don't respond by addressing some of the societal challenges that make it difficult for women to work, many will face financial insecurity and potentially be pushed into poverty.

Stela believes we are at a point where we can align all the players and stakeholders to effect systemic changes to improve the employment prospects of diverse, under-tapped, and underprivileged segments of the population. "Human relationships are the most important thing," she says.

Stela sees an once-in-a-lifetime opportunity to address the root cause of our workforce shortage, skills deficiency, and innovation gap by facilitating a more inclusive workplace and eliminating barriers for women and other minorities.

As you can tell from Stela's journey, she has never feared the unknown and continues to be a champion for change.

1.6   SPACEX: A LESSON IN OVERCOMING FAILURE

These same lessons apply to companies and organizations too. In 2004, the U.S. government passed a law called the Commercial Space Launch Amendment to help with the development of commercial launches. Previously, the government handled human spaceflight and launches through NASA or the Department of Defense. In addition to passing this law, the government also disbursed grants and contracts to companies, entrepreneurs, and innovators to help them create new ways to launch things into space.

One company wanted to try something different that had never been done before. This small start-up wanted to develop a new type of launch vehicle that was reusable and sustainable. Space industry experts said that it was impossible and could not be done. It appeared that they were correct. This company experienced failure after failure in 2006, 2007, 2008, 2015, and 2016.

Yet, they persisted. They learned from their mistakes and improved—and they didn't let the fear of failure stop them. They kept on trying anyway.

In the fall of 2017, SpaceX posted a YouTube video of all of its unsuccessful launches and landings. This video went viral around the world. "How Not to Land an Orbital Rocket Booster," a video with music and some captions, showed a series of "rapid, unscheduled disassemblies" that underscored how important it is to not fear failure but to learn from it. To view the video, go to this link: https://www.youtube.com/watch?v=bvim4rsNHkQ.

Then, in May 2020, SpaceX became the first private company to send NASA astronauts to the International Space Station. This was the first crew to launch from U.S. soil in nearly a decade.

As of this writing, SpaceX is preparing to launch a rocket that is more powerful than the Saturn V rocket that flew Apollo astronauts to the moon. Not only will we return to the moon, but we will move on to Mars thanks to SpaceX's advancements.

Don't fear the unknown—get comfortable being uncomfortable. Do you think Elon Musk relished all of those failures? Watching something you've poured your heart and soul into, not to mention millions and millions of dollars, not work—and with the world watching—had to be pretty uncomfortable!

But they charged forward. The team's tenacity should teach us all to embrace the future and not be held back by fear, anxiety, or ego.

1.7   THE IMPORTANCE OF MENTORING

By now I hope you're seeing some themes emerge to help you make bold moves forward into your desired career—take advantage of every opportunity, try anyway, don't fear the unknown, and grow from failure. I also hope you're energized by stories of people who have persevered to uncover pathways that can change the world. And maybe you've even learned about the space industry and why so many of us are passionate about bringing more women into the global space ecosystem.

This now leads me to perhaps the most crucial element in not only navigating your journey and carving out your story, but enjoying the steps on the way there: mentorship.

At Space Foundation, we've developed what we call the Workforce Development Roadmap (which you can read more about in our eBook, downloadable at https://cie. spacefoundation.org/cie-ebook/). If you think of workforce development as a house, mentorship is its foundation, with the four pillars of the house being awareness, access, training, and connecting; the roof is shaped by the specific types of mentors. Mentorship is so important that it forms the top and the bottom of the house, and that is what I want you to focus on right now.

People are the most important and powerful assets in creating greater access and more opportunities for diversity and gender minorities in the space ecosystem. Mentoring involves building, making, and growing relationships with people who can guide, coach, and champion you as you move across the chapters of your career.

Mentoring is a relationship in which an experienced or more knowledgeable person helps a less experienced or less knowledgeable person. It can take many different forms, but it usually involves giving advice, support, and feedback to assist the mentee in improving their skills, knowledge, and confidence. Mentors provide support in a variety of ways, including personal development, career advancement, and education. They offer advice, encouragement, and resources to enable mentees to achieve their goals and overcome challenges, and there are different types of mentors for different circumstances.

Let's take entrepreneurs as an example. If you're an entrepreneur who needs help in one area, like fine-tuning your positioning statement, you might want to find a "mentor" who knows a lot about writing positioning statements. So, a "mentor" is someone with a lot of skills and knowledge who can help you improve in one area. If you think about a baseball team, a "mentor" could be the batting or pitching coach.

Now a "coach" is someone who helps you develop overall; perhaps your goal is to become an executive leader. Using our entrepreneur as an example once more, many entrepreneurs are the founders of their companies. They were used to doing everything when they initially started their company. However, as their company grows, they transition out of the day-to-day activities of the company and into more strategic activities as the CEO. They might want to improve their executive presence so they are better prepared to meet with investors, stakeholders, and members of the board of directors to get venture and seed funding. They still help their company reach its operational goals and develop its teams, but they want a "coach" to help them improve some of their softer skills. They want to become strategic thinkers and grow personally and professionally. If we use our baseball analogy again, you can think of a "coach" as a baseball coach that focuses on the overall team dynamics and performance. The baseball coach wants the team to become a cohesive unit that wins games.

So, you have mentors who can help you go deep into a subject and coaches who can help you develop as an individual, whether it's leadership, executive presence, mentoring, or something else. And then you have "champions." "Champions" are individuals who can help you position your career. Let's say, as an entrepreneur, you want to take the next step and need a lot of seed funding so you can grow your company. You might be able to find a "champion" investor who brings in several other investors so that you can make a leap with your technology and product development. When we look at our baseball analogy again, think of a "champion" as the general manager or owner of the baseball team. They can position you for progress in your career.

All three types of mentors make your journey richer and easier; they can also help break down barriers you may face and keep you from getting stuck. And they're not as difficult to find as you might think. There are great mentoring programs out there, such as Space4Women, the Women Tech Network, the Space Generation Advisory Council, and the International Astronautical Federation.

And remember, you can be a mentor, coach, or champion to other people as well. You can engage in mentorship at any stage in your career. For example, early-career individuals can mentor college students, college students can mentor high school students, and even high school students can mentor junior high students. This is how we develop our workforce and get more women into STEM careers over the long term.

1.8 TAKING CHANCES

Take chances.

—Sejal Budholiya

Sejal Budholiya is an analysis engineer at Collins Aerospace. She is passionate about aerospace product development, space sustainability, and utilizing design in space for life on Earth. She published four papers and filed for four patents during her undergraduate studies. She's a trained artist, published author, dancer, and entrepreneur. Sejal was a member of the Offsite cohort and BIPOC Scholarship holder. As a designer, she enjoys noticing the unnoticed and designing for those unlike herself. Her superpower is empathy, and along with design research, she tries to solve the most difficult problems we face. She is a mentor, role model, and champion for underrepresented groups.

Sejal has participated in several programs and organizations such as Students for the Exploration and Development of Space (SEDS) India, Space Generation Advisory Council (SGAC), and the Space4Women mentoring program. She believes in the power of volunteering as a form of mentorship, working at a home for children with cancer and with children of prison inmates. She's also created a game to teach visually impaired people how to dance.

Across her various activities, when Sejal would ask various children what they wanted to be, a common refrain was, "What could I possibly be?" She believes in giving second chances and showing new opportunities and possibilities to those who don't feel like they have the privilege to dream. As such, through Neysa, her nonprofit organization, she teaches the performing arts to the underprivileged and empowers their lives through design projects. Sejel is giving back with her passions, and this has allowed her to meet a very diverse group of people. What she has learned is that their joys, pains, and passions are the same as everyone else's—and she has something to offer them, just as she learns from them. In fact, she notes, "Mentoring has been transformational."

Sejal recounts in our Zoom interview (2022d) that she started her journey by looking for role models and people she looked up to who were multidimensional, since she was interested in both STEM and art. She would see female leaders doing really interesting, unique things, and it made her think she could too. Initially, she would enjoy listening to the stories of role models. This led her to eventually participate in formal mentoring programs like Space4Women. Through mentoring, Sejal learned how to handle rejection or get through failure. For example, when she didn't get into an aerospace fellowship program, her mentor was there to support her.

Sejal points out that being a minority can be very difficult in terms of age, gender, and geography. It's helpful to have someone to talk to with professional insight about managing both her personal life and professional goals. She appreciates having that added perspective on how to make life more meaningful as a whole.

Based on her own experiences, particularly with Space4Women, Sejal realized how incredibly important it was to have someone guiding and rooting for her. Without a mentor, many people drop out of the space, technology, and STEM workforce pipelines. She noticed that in addition to the gender gap, there is also a generational gap. The younger generation has a variety of interests and passions, while the generation that has been in the space industry for 20–30 years is very focused on one area based on their experience. Mentors help bridge the gaps.

Sejal actually conducted a global survey and created a white paper about the effects of mentoring programs for gender minorities. She presented her findings at the International Astronautical Congress in 2022. In her recommendations, she advocated for more female role models and for formal mentoring programs to be expanded, as finding a mentor can be a life-changing experience. With the right mentor, more women will be encouraged to join the space workforce. This is integral to workforce development and creating meaningful technologies for the future. In fact, 100% of her survey respondents believed that mentoring improved their professional lives, and many of the survey respondents saw mentoring as a way to "pay it forward" to the next generation.

As I hope you can see from Sejal's example, mentorship is an incredibly valuable resource. It provides the opportunity to learn from the experience and wisdom of others. This is why it is the foundation of the workforce development roadmap. Mentorship removes barriers and creates access points for individuals to find their way into the career of their dreams.

1.9 CLOSING THOUGHTS

Chapters are marked by lessons, whether in life, your favorite novel, or a college textbook. As you prepare for a new proverbial chapter of your life, I hope this physical chapter has offered some helpful guidance.

While I can't speak with each reader individually, this has been my chance to offer mentorship to women who are considering a career in STEM. I want you to be inspired to take action. I want you to be excited by all of the amazing possibilities that exist within the global space ecosystem so that you will consider joining us when you think about careers that involve science, technology, and/or innovation. I want you to know confidently that there is a space for you if you have the courage to take a step forward and turn the page on your chapter. And I want you to believe that you can change not only your own life but the future of life on Earth.

Last but not least, I want you to get out there and DO IT!

ABOUT THE AUTHOR

Shelli Brunswick, COO of Space Foundation, brings a broad perspective and deep vision of the global space ecosystem—from a distinguished career as a space acquisition and program management leader and congressional liaison for the U.S. Air Force, to her current role overseeing Space Foundation's three primary divisions: Center for Innovation and Education, Symposium 365, and Global Alliance. Advocating for space technology innovation, entrepreneurship, diversity, and inclusion, Shelli collaborates with organizations around the world to connect the commercial, government, and educational sectors. Shelli was named the WomenTech Network's 2022 Chief in Tech, 2021 Global Technology Leadership, and 2020 Diversity and Inclusion Officer and Role Model of the Year. Shelli plays an active leadership role with several global organizations, such as: Space4Women, an affiliate of the United Nations Office of Outer Space Affairs; WomenTech Network; World Business Angels Investment Forum (WBAF); G100 Global Chair for Space Technology and Aviation; Global Policy Insights' Global Policy, Diplomacy, and Sustainability (GPODS) Fellowship Program; Tod'Aérs; and Manufacturer's Edge.

BIBLIOGRAPHY

The State of Women. *Amplifying her voice featured speaker: Iroka Chidinma Joy* (2021) *The State of Women*. Available at: https://thestateofwomen.com/amplifying-her-voice-featured-speaker-iroka-chidinma-joy/ (Accessed: 7 January 2023).

Aparna, R. (2023) *10 questions with Sejal Budholiya, Edge of Space*. Available at: https://edgeofspace.in/10-questions-with-sejal-budholiya/ (Accessed: 7 January 2023).

Elsesser, K. (2022) *"Woman" selected the word of 2022 by Dictionary.com—Here's why, Forbes*. Available at: https://www.forbes.com/sites/kimelsesser/2022/12/13/woman-selected-the-word-of-2022-by-dictionarycom-heres-why/?sh=75f16d6662c1&utm_source=newsletter&utm_medium=email&utm_campaign=forbeswomen&cdlcid=618c343d6e1a1d12114458d4 (Accessed: 6 January 2023).

Potter, S. (2020) *NASA astronauts launch from America in test of SpaceX crew dragon, NASA*. Available at: https://www.nasa.gov/press-release/nasa-astronauts-launch-from-america-in-historic-test-flight-of-spacex-crew-dragon (Accessed: 7 January 2023).

Space in Africa (2020) *"To get young Africans into space exploration, we must start by getting an interesting curriculum"—Chidinma Iroka, Space in Africa*. Available at: https://africanews.space/to-get-the-young-africans-to-venture-into-space-exploration-we-must-start-by-getting-an-interesting-curriculum-chidinma-iroka/ (Accessed: 6 January 2023).

The Scully effect (2019) Westcoast Women in Engineering, Science and Technology – Simon Fraser University. Available at: https://www.sfu.ca/wwest/WWEST_blog/the-scully-effect.html (Accessed: 11 January 2023).

The Scully effect: I want to believe in Stem (2021) Geena Davis Institute. Available at: https://seejane.org/research-informs-empowers/the-scully-effect-i-want-to-believe-in-stem/ (Accessed: 7 January 2023).

Zoom Interview (2022a) Dr. Iroka Chidinma Joy, 15 December, Zoom; approved 26 December
    2022 via WhatsApp, 6:55PM.
Zoom Interview (2022b) Priscilla Nowjewski, 29 December, Zoom.
Zoom Interview (2022c) Sejal Budholiya, 22 December, Zoom.
Zoom Interview (2022d) Stela Lupushor, 17 December, Zoom.

2 Be the Change You Want to See

Alex Knight
STEMAZING, United Kingdom

2.1 RECOGNISE YOUR PRIVILEGE AND USE IT TO DRIVE POSITIVE CHANGE

From a very young age, I remember having a strong sense for justice and longing for equity. I grew up with a younger sister who had multiple disabilities and remember raging with anger at school when I witnessed her being bullied by other children. Even as an extremely shy and quiet child, I would fly into a fit of rage at the other kids and try to protect my little sister from this injustice. It must have been the big sister in me, but nobody deserves to be bullied.

I have always considered myself lucky. However, it's only when you have a comparison of what your privilege affords you that you realise the world is flooded with inequality of opportunities. It's the responsibility of those with privilege to recognise where these inequalities are and drive positive change.

My sister and I were born into a life of equal opportunities, but a fluke accident changed it all for her. When I was 3 years old and my sister was only 18 months old, we were in a severe car crash with my mum. The medical professionals didn't think my sister would survive, but she pulled through despite her serious brain injury. My mum was also badly hurt, but somehow, I escaped largely unscathed. I definitely felt survivor guilt, although I didn't recognise it at the time.

As we grew up, I was acutely aware that I had an "unfair" advantage over my sister for anything we both tried to do through our childhood. Any sports or academics came much more easily to me, and I felt guilty for being "naturally better," as it was pure luck that I had the privilege of an easier start in life than her.

I saw her facing challenges that most people would never have to face. Incredibly, she overcame all the odds and now leads an amazingly independent life as an adult. She is still in pain on a daily basis and has very limited use of her right side, trouble remembering things and many other implications of her lasting brain damage. However, she inspires me every day with her positivity, self-empowerment and drive. She is a true role model of what is possible when you put your mind to it and believe in yourself despite what the world is telling you to the contrary.

Looking back, I can see this was the backdrop of my drive to make positive change in the world and empower others to be the change they want to see. However, I lacked the tools to do this until I was much older. Growing up, I was a painfully shy person and struggled socially throughout my entire childhood. I actively avoided the

DOI: 10.1201/9781032679518-3

limelight and went to extreme lengths not to be noticed. My mum recounts the times I would be running my fastest in a sports day race and then look around and realise I was in the lead, so I would pull back to make sure I wasn't the winner! I did not want to do anything to make me stand out!

Now, I can look back with compassion and kindness on 'little shy Alex' who didn't have the tools or understanding about the importance of being seen and heard as the vehicle for helping herself and helping others. I taught myself the hard way, through self-discovery, how powerful this can be. Now, I make it my mission to teach others my signature approach to life – *courage before confidence*! This is especially true for my own children. We talk about being courageous a lot, and I see how this is benefitting them at their young ages.

I'm so grateful to now have the opportunity through my work to help hundreds of other women to develop the self-empowerment tools to be more visible role models and go on to inspire others in a perpetual cycle of empowerment and inspiration. We need more of this in the world, and the change starts with us!

2.2   ENGINEERING OPENS DOORS

I feel that this evolution of the Alex I am today wouldn't have been possible if I hadn't chosen the path of engineering and realised how to overcome the struggles I was facing as a woman in engineering, in the minority. Despite my dad being an engineer, I didn't want to be "just like" him and work on pumps and engines, and didn't see any women engineering role models that I could identify more closely with. So although I knew what engineering was, I only saw a small slither of it and I didn't want that for myself. This was until one of my science teachers at school said to me, "*Alex, engineering opens doors*". This really stuck with me, and I thought, well, maybe I should do engineering and then it will create opportunities and open doors for me. I'm pleased I did, as over the last 20 years, it has done exactly that.

This decision was the start of my personal evolution. From the moment I left my all-girls school and moved out of home to do my mechanical engineering degree at Brunel University London, I made the decision to "fake it 'til you become it" and act like the person I wanted to be. I felt like being a shy person, being the definition of a wallflower, was not serving me, and I could choose to let that go. I have to admit, though, when you start afresh somewhere new where you don't know anyone, this is much easier to do, as nobody has any pre-conceived expectations of who you are. You are free to write your own story without risk of judgement from those who are surprised by the change!

There were still definite obstacles along the way, despite the freedom from my old ties. I felt like I didn't belong at University from day 1, as I didn't get the grades I needed to get in, but was offered a place anyway, on top of comments like "I think you're in the wrong lecture, this is Engineering, English is down the hall". However, I chose to use this as fuel to drive me to succeed. It worked; after a hard-working 4 years I graduated top in my year with a 1st Class Honours Degree and the University Prize for my overall grades, as well as a Prize for my Dissertation. My degree taught me so many things – it opened my eyes to various routes in engineering where I could make a difference. I found what I felt was my niche in engineering – medical engineering, which I chose for my final-year options – where I could use my engineering skills to help people like my sister.

But those 4 years taught me so much more than just engineering. It was the first time I had proved to myself that I could do anything, even if it didn't come naturally to me. I could choose to work hard and just keep going, one step at a time, until I get there. I had proven to myself that I was capable of being an engineer as well as the more visible and vocal person I wanted to be. I had grown my confidence in so many ways, but this was possible only after I was courageous first. You can increase your confidence, but you have to be able to step outside your comfort zone, be brave and then let the evolution begin. I was starting to recognise a key lesson in my life!

2.3 FINDING BELONGING IN COMMUNITY

I used this new level of confidence in myself to embark on a master's in medical engineering at Imperial College in London for a year, and then came the daunting task of applying for jobs. I didn't seem to have much success with job hunting, but to be honest, looking back, this was because I was doing it all wrong. Really, I needed a mentor, someone to help me see the options available and understand how to create opportunities for myself, but I didn't even know the word *mentor* at that time! I did the next best thing and approached the only person I knew involved in the real world of medical engineering and asked him for a job. This opened the door to an interview at a small start-up medical technology company, and I got the job!

I loved it! Getting paid to do interesting work and then have evenings and weekends off was a dream!! However, what I wasn't expecting was that I'd have to endure countless inappropriate comments from the guy who opened the door to this job for me. Things like "Hey, Alex, you should try sitting on a washing machine when it's in its spin cycle – do you know that's the ideal vibration frequency for a female orgasm?" In awkward scenarios like these I would find my old shy persona would emerge and I would just look down speechless hoping the ground would swallow me up. I felt like I just had to brush off these situations as it was all part of being a young woman in a male-dominated environment, right? I didn't know I had the option not to accept vulgar remarks that made my skin crawl. I didn't think I had the right to call it out and let that person know I didn't appreciate his lude form of conversation.

This was pretty much how most of my engineering career progressed. I loved the interesting engineering work, but alongside it was this feeling that I had to endure comments and actions that made me feel like I didn't belong. Unfortunately, lots of other women in STEM fields have gone through similar situations because they were equally ill-prepared. These stories commonly get swept under the rug, and this perpetuates the problem. In most cases I don't believe the perpetrators set out to be patronising or to harass women, but the lack of sharing that situations like this still happen and the impact it can have on women leaves us ill-prepared to deal with it, and ultimately feeling alone and powerless. Whereas if we can collectively raise awareness in work environments, we give the opportunity for everyone to level up their inclusive communication and become allies and advocates for inclusion.

If any of these situations have happened to you in the past, know that you are not alone. I would advise that you find a mentor, and ideally one external to your company. In addition to this, find allies at work and collectively raise the issues to HR, so you are not alone, and together you will be giving your company the opportunity to level up. But if this does not resolve it, know that you have options and never have to endure an environment that doesn't support you. Tap into your mentor's network and join external communities of professional STEM women to find out what their work cultures are like. Use your growing network to explore where the next right step for you could be. You have so much value to add and will find an organisation and team that provide the inclusive culture you need to thrive. There are so many opportunities for women in STEM, so back yourself, be courageous and go find the right next step for you.

For me, my next step involved moving on from that medtech start-up to working in an engineering consultancy where I could learn about many other areas of the vast landscape of engineering. I got to work on amazing projects from submarine integrity improvement, to innovative wheelchair concepts, to working in Thailand on an energy project. As my experience and confidence grew, I'd feel like I was making progress against the underlying current of imposter syndrome, but numerous times another situation would occur where I felt patronised, or recognised for my "looks" or "charm" rather than my technical ability. It was common for people to assume I was in a more junior role than I actually was, and again countless awkward situations arose where inappropriate comments were made, such as "it's nice to have something pretty to look at for a change", "I suppose they needed a Group Leader who was a woman", "why would we want a women's network? What about the men?". Again, back then I still hadn't realised I needed a mentor to talk to about how to handle this, and still didn't have the tools myself to call it out which left me feeling exhausted on many levels wondering whether I was meant to be doing this and whether I truly did deserve to be there.

To be balanced, I did also have examples of supportive bosses, fantastic colleagues and clients. But sadly, there were no women I could see around me in senior technical roles that I felt I could look up to and think, "I want to be like you". There was nobody I could identify with that I felt would be able to help with the specific gender-based challenges I was going through as a woman in the minority. This was a tough time, especially when I had my children and was returning to work after a break feeling like the odd one out more than ever.

Over time, I eventually realised the importance of being part of a supportive external professional community. I joined the Women's Engineering Society and was a member for a number of years where I found peer support. When I moved to a job in London, I became a member of the Women Engineering Society (WES) London Cluster Committee, an incredibly empowering group of women engineers working together to make positive change.

Being in a professional community where I wasn't in the minority, I felt understood for a change, and it gave me the strength that I needed. This inspired me to level up in many aspects of my career and life. Looking back, this was a critical milestone for me, and I think if I hadn't been influenced and inspired by these other women in STEM it would have taken me longer to realise I had the strength within me to truly be the change I wanted to see. A big thank you to you all!

2.4   BEFORE CONFIDENCE

Over the next few years, I felt a passion rapidly growing in me for making a difference where I could apply my natural talents and drive. I'd been mentoring women in STEM internally at work and externally via other schemes for many years by this point and seeing trends in the experiences they were having; lacking self-belief and the tools to help empower themselves. I could see how my experience and the strategies I had taught myself over time were also helping others now. At the same time, I was also doing a lot of volunteering as a STEM ambassador and could see the desperate need for more women in STEM role models for children to open their eyes to what could be possible for them with a career in STEM as well as to challenge gender stereotypes as young as possible. I felt strongly that the diversity and inclusion challenges we have in STEM industries needed to be tackled at the grassroots level.

I wanted to be a part of this diversity movement on a larger scale, so I put myself forward for the WES Board of Trustees. I was voted in and joined the Board in 2018. This was an eye-opening experience, and I gained a lot of insight working at this strategic level to support the growth of gender diversity in engineering in the UK, but I still wanted to do more.

Around the same time, I also made the scary decision to leave a job I had been in for 11 years. This was truly out of my comfort zone, but was actually an incredibly affirming experience. I realised how much value I had to offer to various companies I interviewed with that offered me jobs, and started to recognise that I could ask for what I needed in terms of package and flexibility, then choose the right fit for me. In a lot of ways, I wish I had done that move sooner! I realised I could design the life I wanted with a mix of professional work on a flexible working schedule that enabled my voluntary work in parallel. This satisfied my need to progress my career as well as contribute to the diversity and inclusion agenda.

Joining a London-based team with the mix of diverse people from different countries, backgrounds, cultures and skillsets was a fantastic experience. Also, having a boss that inspired me with his commitment to diversity and inclusion was incredibly empowering. I am grateful for this experience which not only challenged me to grow technically, working on world-leading projects such as the application of IOT predictive analytics to critical infrastructure such as the Forth Road Bridge Crossing in Scotland. The diversity in the team stretched my appreciation of how much I can learn from different people which I could see first-hand led to a step-change in innovation and was recognised by winning multiple Innovation awards. This was a great experience of the benefits of diverse teams which again motivated me to want to be a part of a more diverse and inclusive STEM future.

Using this positive experience as a springboard and taking my own advice – *courage before confidence* – I stepped forward for various other professional opportunities where I could level up and make a difference, such as applying to become a Fellow of the Institution of Mechanical Engineers and gaining a position on the International ISO committee for ISO55000 Asset Management, where I represented the UK at International Meetings to develop the Standard. I also applied for a Visiting Professor grant from the Royal Academy of Engineering which allowed me to do part-time lecturing at Brunel University as a paid position. I knew I wanted to help diverse students with their transition from academia to industry as I had really struggled at that point, and Brunel was the ideal place for me to do that. But none of that would have happened if I hadn't opened my eyes to opportunities and said yes when they were offered! Again, courage before confidence. I initially always felt like an imposter and then realised I had a lot of value to add in all those situations.

2.5 STEMAZING WAS BORN

In 2019 I used this momentum to establish my own "side hustle" – STEMAZING. A not-for-profit social enterprise to empower women in STEM to be confident, visible role models and link this to inspiring the next generation. I wanted to tackle what I saw as a fundamental blocker to equality, diversity and inclusion (EDI) in the workplace and this was a lack of diverse role models in STEM at the grassroots level. It's a long-term solution, but one that is critical for real change in my view.

I had planned to run STEMAZING alongside my normal day job which by this time was as a Technical Director at the London consultancy. But when COVID hit in 2020, I realised I was burning out trying to keep up with all my voluntary commitments, growing my side hustle, my busy day job on top of home-schooling a 5- and a 7-year-old with no end in sight!!

I had to take a serious look at everything I was doing and work out how to get a better balance. After a lot of soul searching, I decided what I really loved and felt a sense of purpose with was STEMAZING. I got the biggest buzz from helping other women in STEM recognise and appreciate their awesomeness and how much value they have to add, as well as empower them with the tools to help themselves build their confidence and become more visible and vocal as role models to inspire our future STEM workforce.

I had started paid mentoring and coaching, as well as paid STEM clubs online for families and schools. I really loved it, knew I was good at it and could see there was a real need for it. It was like a dream – getting paid to do something I absolutely loved that was making a tangible positive impact! However, the salary I was able to pay myself from STEMAZING was nowhere near what I was earning in my 'day job', which was a barrier I needed to overcome!

I had numerous ideas for how I would make it work financially. Throughout 2020, I was developing the concept of the STEMAZING Inspiration Academy – a programme for women in STEM around the World to inspire young children around the World. The programme would initially support the women to shine as visible role models by growing their empowerment and courage, then seamlessly lead into them inspiring young children with our structured proven STEMAZINGKids activities.

I had such a good feeling about this programme as it brought together everything I had learnt through experience and self-development over my life and career, woven into a STEM programme for children that I had already delivered to hundreds of families throughout COVID lockdown, so I knew it would work. It felt so aligned with my values and my drive to create the opportunities for others to be the change we all need to see. However, I still wasn't sure how I was going to make it work financially.

After talking it through with my husband, we agreed we could dramatically change our lifestyle and reduce our outgoings so I didn't need such a high income. We decided we would move house from South East England (London commuter-belt suburb) to North East England, where we could have a significantly smaller mortgage whilst living a life closer to nature in the countryside, which is what we wanted for our family. Again, this felt super scary, but – courage before confidence – we did it anyway! This wasn't the only cutback we made – we even cancelled Netflix!!

We haven't looked back, and we now both work from home from our cottage in the Northumberland National Park, where I get to focus solely on STEMAZING and now run two rounds of the STEMAZING Inspiration Academy programme each year, have a STEMAZINGWomen Membership and deliver STEMAZINGKids initiatives. We have a model where STEMAZING Partners who are aligned with our values of inspiration and inclusion in STEM contribute to funding our work, and benefit by being involved in our initiatives. It's amazing how many people in the world want to be a part of something empowering and inspiring like this, so I make it as easy as I can for them! It wouldn't be possible without their support – it is a team effort!

Through our programmes, at the time of writing, we have empowered over 350 women in STEM to be more confident visible role models around the World, and they have collectively delivered over 100,000 STEMAZINGKids activities for 7- to 9-year-old children in deprived areas, from all around the UK (Cornwall to the Outer Hebrides) to Kenya and Ghana.

I feel incredibly grateful to be in a position to *be the change I want to see*, to be a role model for others for what is possible when you choose to live courageously, to be able to live true to my values and have the outlet for my innate need to make a positive difference.

My sister also now lives in the North of England, and she continues to inspire me to be a force for positive change in the world, to use my privilege to help empower others and create the foundations for greater equity and inclusion in the world. This is only the start!

For more information about STEMAZING, visit www.stemazing.co.uk.

ABOUT THE AUTHOR

Professor Alexandra Knight, CEng FIMechE FWES is a Chartered Mechanical Engineer, a Fellow of the Institution of Mechanical Engineers and a Fellow of the Women's Engineering Society. She has had a varied career working in industry for almost 20 years. Alex's most recent job in Industry was as Technical Director for an Asset Management and Data Science Consultancy, where she led a number of projects focussed on enabling Digital Asset Management in Infrastructure. Alex was also a member of the ISO UK Mirror Committee for Asset Management, representing the UK at International Meetings to improve ISO 55000, the International Standard for Asset Management.

Alex is a passionate advocate of diversity and inclusion in engineering. In 2019 Alex founded STEMAZING – a social enterprise dedicated to inspiration and inclusion in STEM. Through the organisation's two key themes – STEMAZING Women and STEMAZINGKids – Alex leads several not-for-profit initiatives to support women in STEM to shine as visible role models and inspire our future generations of innovators and problem-solvers. As of 2023 they have empowered over 350 women to be more confident visible role models and have collectively delivered over 100,000 STEMAZINGKids experiments.

3 I #DERRtobedifferent, I Dare to Be Me

Ludmilla Derr
Elite Experts Conferences Ltd, Switzerland

There are many ways to success and happiness, there are also many definitions of success and happiness. It is all about finding your own unique way. #DERRtobedifferent, dare to be you. I've been thinking for a while about how to start my chapter. Chronologically? With a jump to the present and then gradually explaining how I got there? Both beginnings have a justification. Personally, I find it more exciting to be fascinated by the actual state of a person and only then to puzzle out about how the way to get there was. It is motivating for me to see the end result in advance.

However, my path has had a twist at one point, so if you don't know the previous very straight path, you can't even comprehend what a significant turning point it was. I will do both – a mix of chronological run-through and several jumps in time. After all, we don't want it to be boring, right?

3.1 THE PATH TO MY OWN TRUE IDENTITY – A CHRONOLOGICAL QUICK RUN-THROUGH

I was born in Kazakhstan, moved to Germany at the age of 14, had school education, high school diploma, the most unusual first job at 17, attained a degree in chemistry studies as well as a PhD in material science. I moved to Switzerland at 31 for a great job at a Swiss company for global projects in the automotive industry. After two years gaining a CAS (certificate of advanced studies) in event management parallel to the job and the birth of my son, I quit the job and started my own company in technical marketing and event management. Five months later, the Covid pandemic and global 'difficult economic situation' began.

Let me be realistic – it could have been over for my company sooner than you think. But giving up was never an option. In contrast, building sustainable networks, generating satisfied customer projects, testing and expanding new concepts, changing and optimising the range of services – all this was what made it possible for such a young company to survive the pandemic. I can even say that the way the company is well positioned now was largely only possible because of the pandemic.

That was the chronological quick run-through. For those who think that this path was full of roses and candies, I have to disappoint. It is not a modern fairy tale. But it's not all bad and dark either. It's a colourful mix of winning and learning, winning and learning, just as colourful as life itself. It's real life, and as we all know, life writes the best stories.

DOI: 10.1201/9781032679518-4

3.2   TURNING POINT AND WALKING IN THE FOG

It's late summer 2017, early in the morning, the sun has just risen. I am on the beach in Vernazza, Italy, on the Ligurian Sea. I am sitting on a flat rock that was big enough to place a whole picnic blanket and still have extra space to do my morning yoga exercises. The yoga exercises are now finished. Namaste – a new day can begin. A day on which I write down for the first time in my morning pages that I would like to completely reorient myself professionally. The vision is just beginning to take shape. The way there – stony and hard?

No, much worse – the way there is not clear at all at that time. When I think back now, it seems almost naive how I thought back then. Or is that true? On the other hand, if you want to get from A to Z, who said it would be bad to focus only on A, B and C at first? Sometimes, if we knew all the difficulties from the beginning, we might not even start. And that would be a pity, not only for ourselves, but also for all the people we would never help, because we never start.

Lesson learned:

> It is good and important to have a direction, but the whole vision may also develop over time. That shouldn't stop you from getting started. It's like walking in the fog – the next step becomes visible as you walk.

Just to give you an idea of how much I wasn't yet committed to what it was going to be, I'll give a few examples. I had the feeling – probably this is some remainder of the 'good girl' syndrome – that if I start something new, then I have to learn it first. So again, back to university. I didn't want a whole degree. But at least a CAS, i.e., a kind of mini study, which you can put on top of some years of work experience.

I found out about the various CAS studies that were available in Switzerland. I looked at marketing, international relations and multicultural understanding, diplomacy and business psychology. Honestly, when I retire someday, I'll do another degree program – there's just so much exciting stuff to learn in this world. I remember flirting with a CAS study in foreign affairs and applied diplomacy for a long time. The content of the program was super exciting, and on top of that, the modules would always take place in different cool places, like New York, Beijing, London and many more. Learning and travelling always go together for me, so it seemed very attractive. Then, I saw the CAS Event Management, where marketing, project management, leadership, creativity and so much more came together, which intrigued me. No, it didn't take place abroad in cool places. Instead, each module would take place in a different location in Switzerland. Why? Because you'd be live at the coolest world-class events, combining theory with practice behind the scenes.

3.3   HERE AND NOW

At this point, I would like to take a brief jump in time, to the present. I run my own company, Elite Experts Conferences, based in Switzerland. We help our international clients execute exciting technical marketing projects. Sometimes we give the stage for that and present the most exciting people in the automotive industry, and sometimes I get invited to do technical interviews at external conferences. We are also active behind the scenes, helping to set up conferences and panel discussions, organising keynote speakers and facilitating collaborations.

But the most exciting branch of our business, and also the one that gives me back the most emotionally, is the trainings and one-on-one coaching in technical marketing. I personally coach some C-level personalities in thought leadership when it comes to how they should position themselves on LinkedIn, and I coach whole groups of employees in the technical departments, but also in HR, employer branding and even marketing teams. We work with large world-renowned companies, and we work with start-ups and scale-ups as well.

Four years ago, I myself went down this path of my own thought leadership, public positioning and authenticity. Now, I see a lot of people going this way as well. It is a scary path when you first start out. However, a self-confident combination of expertise and personality always works – not when it is only presented to the outside world, but when it is truly lived. Therefore, when I start individual coaching on thought leadership with the leaders of the tech world, I always start with finding their true values that guide them through life. No one should ever try to fulfil dreams that are not their own dreams. It is really great to see these personal transformations and to see the shining eyes of my 'students' when they also discover this magic of thought leadership for themselves. That's what truly gives me the greatest value, seeing them uncover their magic.

The vision has also become clearer. I want to make technical marketing more personal, more human and more exciting. I want to bring together the most exciting, smartest and inspiring people in the automotive industry and foster collaborations. I want to make the technical world of knowledge a bit more tangible, understandable and sustainable. I want to help people who have a lot of valuable things to say become more visible. I want to transform many lives for the better.

So how does someone who has a degree in chemistry and a PhD in materials science end up venturing into the world of marketing and teaching others how to do the same?

3.4   KAZAKHSTAN – WHERE MY PATH BEGAN

Let's go to the beginning of the story. We will travel, and we will travel twice – once in time and once in geography. I was born on a March day in 1984 in Kazakhstan to a Russian-German family. Or, if you want to be very precise, I was born in the USSR, because Kazakhstan became independent only when I started school in 1991. There are studies that show that certain factors in life, especially in childhood, have a strong influence on the chances of success later in life. Well, I can really consider myself lucky in many areas.

- *First Advantage*: I grew up in a happy family, where parents love each other, where children – my older sister and I – receive love, attention and support. I don't want to hide how important my family is to me or how big of a part they play in my success. Because it is really worth its weight in gold to have loving and supporting parents that are still a happy couple together.

It's incredible – my parents have been a couple for 48 years and married for 42 of them. Thank you, Mom and Dad, for giving me a simply amazing feeling of a loving home and the famous roots and wings!

• *Second Advantage*: Now this may sound unbelievable to someone, but being born in Kazakhstan as one of the countries of the USSR can be an advantage. Hard to believe? Well, there are even two points of advantage. It was normal in the USSR for both men and women to work in professions that are considered to be male professions in Europe and beyond. Besides, mathematics and science were always the top subjects in school education at that time, and that's where most emphasis was placed. That corresponded perfectly with my own interests: mathematics, chemistry and physics. I did not experience any cultural or social limitations when I was interested in these technical subjects as a girl at that time, nor did I develop any false beliefs inside myself that I could not do something or not do it well enough because I was a girl.

This is how my path into the technical world began in Kazakhstan. I was a bright and curious child. My parents, especially my mother, always supported me. Thanks to my mom, who always gave and still gives good examples with countless books that she reads, I was able to read already at the age of four. Perhaps some of my interest in technical subjects can also be attributed to the adventure novels of Jules Verne. These were pretty much family reading in our house. There are such wonderful moments from childhood when the four of us went camping. My dad would be out on the boat fishing, and the ladies of the family would be relaxing on the shore, and one of us – my mom, my sister or me – would then read aloud from a Jules Verne novel. When one got tired, the other would continue reading. Nowadays, there are audio books for that, but in the old days we read to each other.

Two more aspects should be known about schooling in the post-Soviet period. In retrospect, these also bring two great advantages and made a good school for life: competitions and presentation skills. In my school days there were many competitions on all kinds of topics, held between schools, cities, regions or even nationwide. I was allowed to participate in such events at a very early age.

I don't know if you are born with it or can train for a lot of things, but after seven years of schooling in Kazakhstan, I could handle arithmetic or reading under time pressure and any competitive situation pretty well. We even practiced debating – I found that totally fascinating. The second aspect was that you had to do a lot of recitation in front of the whole class or present something on the blackboard. I loved that, so I was also allowed to perform on stage at inter-class celebrations with mixed groups or with some kind of theatre or dance performance.

These early stage experiences shaped me in such a way that I have no reservations at all about presenting something in front of a large audience. It is exciting to observe how this has changed over time. I still love the stage, but above all, I love to lead technical discussions as a host, as a moderator with exciting personalities from the automotive industry. Providing a pleasant atmosphere and getting the best of knowledge and life experience out of each panellist became much more important than giving my own presentations. That's the transformation of time.

I summarise this first phase of my life: a happy childhood, an extremely curious and intellectually inquisitive child who found interest in pretty much everything, but first and foremost in mathematics, chemistry and physics. In addition, let's keep in mind that through neither my home nor school nor society would I develop any limiting beliefs that as a woman you can't or shouldn't do something.

3.5   MOVING TO GERMANY – NEW BEGINNINGS AND MAKING THE 'IMPOSSIBLE' POSSIBLE

In this part I would like to tell you a very personal story about daring goals and why it is often possible to achieve the 'impossible'! I'll take you on a journey that goes back to 1998.

But let's start from the beginning. In December 2020, I welcomed Yann Vincent, CEO of Automotive Cells Company, as an exciting guest on the Elite Experts Conferences Technology podcast. In the last part of the interview, where I always talk to our top guests from the tech world about their journey, their sources of inspiration, their challenges and their failures, he said this one sentence that was very remarkable for me personally. When I asked him what his best lessons in leadership were, Yann Vincent replied the following: *'Set yourself very demanding targets – that's the best way to respect yourself.'*

It resonated so perfectly with my own attitude, which has been with me all my life, that it got me thinking. I always set high ambitious goals for myself – very important – always linked back to MY values and MY priorities. But how did it all start? What is the magic, the charm and the immeasurable power behind high goals? And more importantly, why should any of us set challenging and even 'impossible' goals?

This part of the story could also be titled 'If I can do it, so can you!' I want more people to benefit from this experience, from those life lessons. First of all, I do believe in new beginnings, really challenging new beginnings! What do you associate with new beginnings, especially when you think of your younger years? Is it starting school? A relocation? A change of school? Or when you started learning a new foreign language?

My first formative experience with new beginnings had been a little bit of everything. As you already know from the beginning of the story, I was born and grew up in Asia, in the beautiful north of Kazakhstan, in a town called Kokshetau. Now let's travel in time.

It was summer 1998. I was 14 years old when my family moved to Germany. All at once – a new language, a new school, a new country, a new social culture! If being a teenager is not challenging enough, then a move like this turns your life upside down!

I remember sitting in the back seat of a taxi on July 22, 1998, with my family on the way to the airport in Astana, where our flight to Hanover, Germany, was to depart. I was looking through the rear window at the fields, steppe and forests moving away from us. In a figurative sense, I looked back on my life so far. A life that would soon no longer exist in this form. I still remember how I inwardly let go and prepared myself for a new beginning without any expectations. I decided not to worry about how everything would turn out, but to look ahead with confidence. My dog, my school friends, my familiar home and my 'image' of being an intelligent and inquisitive child stayed behind.

In the fall of that same year, the new school year started in Germany and with it something like a new life began for me – my second enrolment in school, so to speak. Until then, I had grown up at home with Russian as my only mother tongue despite the fact that my father is German. My grandfather was the only one who actively spoke the old German language. At school in Kazakhstan, I learned two more languages: German for three years, and Kazakh for seven years. But the German lessons were mainly translating texts and learning grammar – a bit like Latin lessons. Only that Latin is a dead language, but German is alive.

With this limited basic knowledge, the changeover was quite overwhelming for me. Do you know the feeling of not being able to distinguish in a foreign language where words begin and where they end? And everyone spoke so fast that my school German couldn't help me at all. In the beginning, I couldn't get any grades in German, politics, history, geography, etc., because a fair grade wouldn't be possible due to the lack of language skills.

Now, let's jump in time again. After a single year at school in Germany, I perfected my German language skills to the point where I spoke grammatically absolutely correct and even *accent-free*! Although the supposed language experts thought it was quite impossible that one could speak accent-free when entering a new foreign language at the age of 14. Yet, that is exactly what happened! Not only do I speak German, but I *feel* this language and culture, I can dream in this language – it has truly become my second mother tongue. If you don't believe me, listen to the Elite Experts Conferences Technology podcast – even in English I already speak with a German accent and not Russian. The story then went even further: after three years at this school, I graduated with the best degree in the entire history of this school! I still hold the record.

Sounds like a perfect, beautiful fairy tale? Everything very simple, all rosy and perfect?

By far not ...

It was difficult, challenging, an *extreme* amount of work, many trials, many mistakes, falling down and getting up again, perseverance and, again, perseverance! That is where I learned grit! It was also anything but easy for my sister and me to be the only two students with an immigrant background in the entire school. There were even a few kids in our school who tried to bully us. They tried, and they failed. It's impossible to bully somebody who is self-confident by nature and doesn't need social approval from a group I never wanted to belong to in the first place. However, that was also formative, and I have to say it prepared me very well for the fact that I now have no problem belonging to any minority anywhere and standing up for myself and others.

But it was also great fun and joy to approach something completely new and to be allowed to make mistakes. And of course, I had great teachers who supported me. I was lucky to have had really fabulous teachers who all were passionate about their subject. German, chemistry and math teachers are the personalities that really shaped my path back then. Thank you!!!

My German teacher in particular was super remarkable. She was also the English teacher in parallel. In English class, I was present but did not participate. However,

I was allowed to use the time to do more supporting exercises in German. And so the teacher observed me a few times, how I joined in the English class after I had finished my German exercises.

I remember very well that exactly at that time, we had a class trip to the ice skating rink. Until then, I had never stood on ice skates, let alone ice skate. But it was explained to me and I tried it out. Sure, I fell down several times, but I always got up and tried again with a higher enthusiasm. After a while I was able to hold myself on the skates quite adequately, and I had a lot of fun doing it! Then my teacher came up to me, praised my way of not giving up and learning fast and offered, if I wanted, to teach me extra English in her spare time. Of course I said yes!!!

That was amazing, we sat together in free hours or after class and caught up on English grammar step by step until I joined the class quite regularly. The fact that I am now writing this book chapter in English, I owe in the end very much to my brilliant teacher and my perseverance!

Of course, many new beginnings in my life followed after that, and of course some more new foreign languages were added – you can never speak enough languages. It's no joke; one day I would like to speak all the major languages of this world. Maybe not accent-free, but still quite reasonable.

Life is long and meaningful if you know how to use it.

What are my lessons learned?

- Before you start running, you have to define the goal very precisely and it has to be YOUR goal, based on YOUR desires and YOUR priorities. Life is precious; don't waste time to reach other people's goals. L-I-V-E Y-O-U-R L-I-F-E!
- Do what you love and what your heart burns for! It is a priceless feeling to love your work and to have fun with it every day! P-R-I-C-E-L-E-S-S!
- Always believe that you can do *much* more than you think you can do sitting in your comfort zone. You can learn to constantly move outside of your comfort zone and feel good about it. It gets easier every time!
- Don't be afraid to make mistakes. Mistakes are part of the game. It is only crucial if and how fast you learn from it. If you knew how many mistakes I made in German at the beginning. It is also okay to be down after a defeat. Give yourself some time, but then get back up and make the next try, and the next try, and if necessary the next try.
- It is okay to be against the opinion of others if you are absolutely convinced that you are doing the right thing. You can feel if decisions are right on the inside – TRUST THAT FEELING!
- If someone tells you something is impossible, it is impossible, but FOR THEM, NOT FOR YOU! It can't be said enough, never let limited knowledge, limited imagination or limited perseverance of others get you down!
- Create from the beginning a great network of positive, courageous, inspiring people who have a similar inner fire in them like you do.
- This is probably the MOST IMPORTANT LESSON of all: if you can do something – teach it to others, if you have something – share it. Always help as many people as possible to get ahead with their goals.

I hope that you, dear reader, will also be infected by this positive energy, and that you will also set yourself great goals and ACT! Here's to an inspiring new beginning, to your goals succeeding, to the magic of making the 'impossible' possible!

Let me close this part with a great quote from one of the most famous philosophical poems by Hermann Hesse: 'Jedem Anfang wohnt ein Zauber inne,' which in English means *'There is magic in every beginning.'*

3.6   MY VERY FIRST JOB IN TECH AND THE IMPORTANT ROLE OF MENTORING IN MY CAREER

When talking about my path I have to emphasise how important mentoring was in my career. Let me tell you about my first mentor. Do you know those Hollywood movies where the leading actor appears at exactly the right place and time to get a chance – usually THE chance of a lifetime?

Totally unrealistic? Not in my case: life writes the best stories.

But see for yourself! I was 17 at the time, was in the 11th grade at a Gymnasium – German high school – in Braunschweig/Germany. I was in a train, just on my way from a student exchange in Italy. We were full of energy, in a good mood and, well, very loud. At one of the intermediate stops a business traveller came in and sat down with our group because the poor guy made a reservation right in the middle of our group. I think he even rolled his eyes a few times, because our loud conversations and laughter certainly gave him the 'peace' he needed for work.

One of my classmates was about to involve me in a heated discussion where a lot of rhetoric was needed. It was about my home country Kazakhstan and about the integration of German late repatriates – this is the official term for it – in Germany, for those who come from Kazakhstan, Russia and many other countries of the former USSR. At some point, the businessman must have started listening to us. And then he handed me his business card and said that he would offer me a job at the Fraunhofer Institute in the same city where my school was. I could call him to set up an interview. He would be looking for someone who could do technical Russian-German translations.

Years later, he told me that through my way of discussion and argumentation he recognized my potential and wanted to give me a chance – and then see what I would do with it. Well, the point was this: I wasn't looking for a job. After all I was well occupied with high school and preparations for the exams. I had been in Germany for only three years and was working hard to perfect my German. But the curiosity to get a taste of the research triumphed. So, I did something about it: I called, went for an interview and got the job at the Fraunhofer Institute WKI, and the businessman became a kind of mentor for me for the next eight years.

For me it was always clear that I would go in the direction of natural sciences and that I would start studying after high school. This quickly became apparent in the job with the technical translations, and after the first three months I was allowed to work in the research laboratory. I was thrilled!!! Of course, it required so many conditions because of occupational safety, etc., to let a high school student (!) without proper education (!!!) do chemical research in a laboratory in Germany. Imagine the bureaucratic effort! But for my mentor this was no obstacle at all – he supported and promoted me where he could. In the end, I worked there for three years parallel to high school, then five years parallel to chemistry studies – eight years of practical research experience in total at the end of my studies.

It was always so great to see everything that you learn in theory at school and at university put in practice or to see that something doesn't work in practice. My mentor always wanted me to see how real research projects are started, managed and completed. I was allowed to be present at all the important meetings and made myself useful there – I wrote the minutes of the meeting. After the meetings, I always analysed what was said with my mentor, talked about strategies and discussed the motives of the project partners. I was also always allowed to see the fruits of our work and was invited to come to the fairs when our results were presented.

It was great how my mentor dealt with my new ideas in the lab. He always (a) listened and took them seriously and (b) allowed me to try them out on my own on a small scale to see if they had potential. At that time, this way of dealing with my ideas was absolutely natural for me – only later did I understand that he was very unique and that these are really cool qualities of a real leader.

With the years of experience in the lab and in the technical field, I already did talent scouting for my mentor's department as a chemistry student. After all, I knew the talented chemistry students from my year group very well from our lab internships. In this way, some of my fellow students have also found their way into practice, and I am very pleased that many of them have remained loyal to the Fraunhofer Institute. Sharing good things is extremely important!

What have I learned from this chance for my life?

- Chances are everywhere – you have to see them, seize them, use them and also give others a chance.
- Sometimes luck, coincidence, fate or karma – depending on your beliefs – decide that you get a chance, but if and what you do with it is entirely up to you.
- Over and over again: 'who knows the goal, finds the way.'
- You can also create a job 'on the job' according to your preferences and talents, so the job with technical Russian-German translations can be turned into a laboratory job in the applied research laboratory.

3.7   THE STUDY OF CHEMISTRY

My love for chemistry was pretty much love at first sight. It was so logical and understandable for me right from the start. Yet exciting enough that there was always something new to learn. I already knew from about the eighth grade that I would study chemistry after high school.

My first job at the Fraunhofer Institute only reinforced it. Not only was I able to perform real useful lab experiments myself, but I also got to see the job of a 'normal' chemist in applied research firsthand. You may be surprised, but I knew right away: THAT is NOT what I would want to be. So, to sum up: Study of chemistry – yes, but afterwards, the job of a 'normal' chemist – no thanks. And I already knew that BEFORE my chemistry studies.

From my parents' house I had not been given any instructions, recommendations or directions at all. My parents experienced my curiosity and determination early enough, so they never had to check my homework, but especially my mom would rather remind me to take more breaks. Therefore, my parents completely trusted that I would choose the right path. Still, my mom would ask from time to time if I wouldn't rather like to study foreign languages, because chemistry is so dangerous and I'm so good at learning languages. My answer to that was always: 'Knowing only foreign languages is like a package without content. I need the content first, the technical content and later I can still talk about it in all kinds of languages.'

In my studies, I encountered for the first time certain prejudices and saw that women are discriminated in certain male domains. Although my study program consisted pretty much of 50% female and 50% male students, it was mainly due to the fact that in the field of food chemistry, mostly women were enrolled. My major subject was physical chemistry, which is a pretty male-dominant area.

Anyway, chemistry studies are structured in such a way that anyone who is not serious about it is screened out at the beginning regardless of gender. Many students quit during the first semester. On one hand, it is a surprise for many that half of the day is spent in the lab. In the morning there are lectures. After lunch, there are lab courses. A lot of lab courses and very long days. For every single sub-subject of chemistry, there is a practical lab course. If you don't love it, you're out of it pretty quickly.

Then, there is also the fact that a lot of mathematics and physics were necessary. Whether you can do math or not also determines whether you can keep up with the material and be among the best. I had math and chemistry as my major subjects at school, so I was one of the few women who had a good chance in these subjects, which are classically considered male disciplines.

Not surprisingly, I was the only girl in a clique of five boys who were all strong in math and chemistry. I remember that at the beginning it was surprising for my male fellow students to see someone of equal ability in a girl, but you can't do anything about ability. The facts spoke for themselves. In addition, since I am self-confident by nature, I was always able to defend myself verbally in a skilful, humorous and intelligent way. Sometimes you have to beat the 'enemy' with his own weapons.

After the 'balance of power,' so to speak, was clarified, we became a strong group of friends who stuck together well. Especially in the third semester, shortly before the intermediate diploma, we were all heavily tested. The number of lab courses, protocols and oral exams were raised to another dimension. In addition, we had a written exam in organic chemistry every four weeks.

If we failed two exams, we would have to repeat the whole course. If you failed one exam, you had the option to repeat only the exam – if you then passed, the whole course was considered passed. The stress level was immeasurable, free time was not only zero for us, it was in the minus. Even more students left the university. My group stuck together, and together we got through the difficulties until we all got our diplomas.

If I had to give someone recommendations now on how to get through a fiercely difficult task well, it would be this: form a community as early as possible with 'your people' who have similar goals, who are similarly ambitious and who have similar values. That's what can be critical. Without that mental support, without that knowing that my friends are overcoming the same difficulties, I wouldn't have gotten through it.

Knowledge alone will get you far, but it has limits. Strong community gets you even further, even when you're given tasks that are impossible to complete within the constraints of time. That's why I always emphasise: help everyone on your path who is going down a similar path and may even be a few steps behind and please help regardless of gender.

3.8   PHD IN MATERIALS SCIENCE – THE IMPORTANCE OF PATIENCE AND PERSEVERANCE, AS WELL AS HEALTHY HABITS AND NETWORKING

Most chemists decide in the course of their studies for or against a subsequent PhD. In fact, the majority actually decide in favour of it. In my case, it was a little different from the beginning. I knew that I wanted to do a PhD even before I started studying chemistry. In other words, I already knew at the beginning of my studies that I would determine the course of my life for the next eight to nine years: five years of studies and three to four years of PhD.

Why was I able to say that so precisely? Well, through my job at the Fraunhofer Institute I got to know other research assistants and doctoral students and exchanged ideas with them. While I was still a high school student, I was always able to see my future by seeing other university students already going that path. I also saw the ups and downs of that future closely enough, but still with a safe distance.

So, it was clear – I would do a PhD. What was also clear to me was that there was no way I was going to do my dissertation at the same Fraunhofer Institute where I had worked the whole time in addition to school and my studies. I worked there for a total of 8 years. Thematically, it would have been so easy to get into a PhD topic there but exactly that made me feel dizzy from boredom.

No, I didn't want easy or boring. I wanted new and exciting!

So, in the final months of my diploma thesis, I went looking for the most exciting topics for dissertations. One topic immediately caught my eye because it was so interdisciplinary. So interdisciplinary, in fact, that I would even change faculties: engineering instead of natural sciences. In other words, it would even show in the title. A 'normal' chemist has the Dr. rer. nat. degree; I am actually Dr. Ing. How fitting for my #DERRtobedifferent life motto! But the exciting thing is, of course, not the title you get for it, but the topic. It was a topic at the intersection of materials science, nanotechnology, physics, biophysics, biochemistry and, of course, chemistry itself.

Did it scare me that I would have to learn some areas completely from scratch for my job? Actually, not at all! I like to go behind the books, do fundamental research on a topic that is completely new to me; and I love to learn new things – preferably every day. I'm more concerned about the opposite, when I realize that I'm no longer learning anything new, or that I'm only learning something new sporadically. I don't like stagnation. Or, if one day I wake up in the morning and I don't have my drive and curiosity anymore – that would be even worse!

Did everything work well from the start? Not at all. Actually, something very basic didn't work right in the beginning. I had to measure the effect of adsorption on the activity of enzymes after they adsorbed on certain nanoparticles. But my enzymes did not want to adsorb on the nanoparticles. It took me a while to find proper test systems and to develop suitable lab assays. All in all, my PhD reminded me of the importance of patience and perseverance, as well as healthy habits and networking. While patience and perseverance are self-explanatory, the other two factors I would like to explain a bit.

What few know during this time I came very close to a burnout. Maybe burnout is too strong for that, but at least I came close to it, and it was no longer tolerable for me. In hindsight, I see it like a natural learning process that I would have to go through once for hopefully my whole life.

But first – how did it come to that? From my school days and my studies, I was used to a lot of learning. And it was fun. Maybe with a few difficulties and obstacles, but in the end everything worked out well. That was exactly the difference. It didn't want to work out well. My basic systems in my dissertation, which I worked on from the beginning, didn't want to work. So more literature research, more tests, more work. Still didn't work. I was making the rounds. More literature research, more tests, more work. The only method, the only way to success I knew so far – more work – suddenly didn't work anymore. We all know that too much work makes you overtired and prevents you from thinking clearly, let alone creatively. A vicious circle.

I started working over the weekends – the technical challenges didn't let me go even in my little free time, and the free time became less and less. I cancelled more and more meetings with friends, my sports program consisted only of cycling – from home to the institute and back again.

For such high performers, among whom I counted myself so far, it is also an overcoming to admit that something is wrong or that you are overwhelmed with something. My partner was my confidant with whom I could talk about it, which is also extremely important. I knew that if I continued like this, my health would be damaged.

The turning point for me came when, despite having a cold, I was struggling through a week of work instead of giving myself rest and time to get better. That was the final warning sign for me and it flipped the switch in my mind. A new project was born: my health and well-being as a priority.

The irony in the story was that I also approached it much like a new work project. First of all, literature research. I remember that I went to the city library of Bremen on Saturday. Unusual in our time with internet, isn't it? But I wanted to be far away from all computers and screens. Besides, I associate libraries with nice leisure activities since my childhood. I am even writing these lines in a library.

I first researched everything about what stress is, how it is triggered, how to deal with it and what mechanisms there are to prevent burnout. I started to learn and apply all this – self-therapy, so to speak. I am not saying that this method is the only right one, but it was the right method for me. It was a way of self-awareness, balance, setting boundaries and priorities, a way I still walk. I managed to pull myself out of this vicious circle, to find a balance, to bring creativity and playfulness back into my life.

All this has also led to the fact that I was able to find practical solutions in my dissertation and successfully complete my work. In the meantime, I am also very good at recognising as soon as the first signs of stress appear and can quickly counteract them. Having experienced that extreme level of stress also helps me quite well in the present when I coach C-level people in thought leadership, because understanding what some leaders go through on a daily basis helps to find a common denominator in communication.

But now I ask practically: did I suddenly have time for my partner, my friends, sports and free time? Yes, because I defined my priorities. During this time, I started with traditional Kung Fu, which became a total passion. I did a windsurfing course, swimming lessons and dance classes, spent a month in Miami doing a language course and a month travelling around in China training Kung Fu, and my Sunday brunches with friends are still legendary. I remember the last few months of writing my dissertation before submission, which many find extremely stressful, which was very peaceful for me. All thanks to my newly calibrated healthy habits.

I knew how to get into the writing flow, how to keep stress low, and what small doable steps I should take in a day to have good results tomorrow. Two things were extremely important during this time: almost daily one-hour walks after lunch with my partner, who was also working on his own project at the time, and working according to the principles of the book *The 7 Habits of Highly Effective People* by Stephen Covey, which I highly recommend.

There is one more factor that played an important role in my PhD time, and that is the network. In my institute, my PhD topic was a niche topic – too interdisciplinary. This led to the fact that I could not really exchange scientific ideas with anyone. But when I had already been working on my topic for a few months, a project between three institutes started and my topic was in the middle of it. Since then, about every two months, I would present my results in front of three professors – one of them my PhD supervisor – several post docs and other PhD students. It might sound like a horror scenario to some. But for me it was extremely valuable experience that I could grow from.

When you have a high-profile audience, you quickly learn how to present your data well, how to explain your hypotheses plausibly to others and how to defend your opinion when you can back it up with facts. I already have it in me, character-wise, that when I have a choice of 'fight or flight,' I go for the fight. This is perhaps due to the competitive situations in Kazakhstan since the very early years.

However, what was made possible by this new constellation is that I had really great discussion partners with whom I could have scientific conversations. One in particular stood out: the professor from biochemistry. He was so kind to invite me to his lab to work together with his PhD students. There, I got my own lab place and felt extremely welcome. Not on paper, but in practice, this professor became my real PhD supervisor, a real scientific advisor who accompanied me until I successfully defended my thesis.

This shows once again how important it is to have the right people around you, how important it is to seek and find exchange. It also happens that people who are assigned to us as our mentors do not have the appropriate knowledge or leadership qualities that we need to move forward. It is up to us not to fall into the victim role and complain about how unfair life is, but to look for suitable mentors outside our own institution.

That's why I always tell anyone who asks me for the recipe for a successful dissertation that it's these four factors: patience, perseverance, healthy habits and networking.

3.9   MOVING TO SWITZERLAND

Can you be a rational-minded person and have highly developed emotional intuition at the same time?

Yes, it is possible!

But first let me share with you how I moved to Switzerland on June 1, 2015. Actually, the story even starts in 2012. I was a PhD student at the University of Bremen in Germany – somewhere in the middle of the PhD process. Hypothesis – tests – validation – writing – repeat. Anyone who has been through this in natural science topics knows what I am talking about.

In May 2012, I came to Switzerland for the first time to a Gordon Research conference in Les Diablerets in the French-speaking part of Switzerland. After the five days at the conference, I added a long weekend over the Whitsun holidays with friends in Zurich. I remember standing on a wooden footbridge on Lake Zurich in the glorious sunshine, looking out across the horizon, and I had this special feeling of 'Wow, Zurich is so familiar to me and this city means something to me – how is that possible?!? This is my first time here, but I am at home here.' So much for the emotional level in that situation.

Several turns and decisions in my life followed, such as going into industry or pursuing a professorship, post doc position – yes or no etc. But what accompanied me through all of this was the vision of this foreign, yet familiar city, of the beautiful mountains and that unbelievable feeling of being at home here.

I needed three years to fit that vision in my life and to reach that destination, so to speak.

Lessons learned:

- Was that easy? No, nothing is easy that is worth achieving. Think about it the next time you have difficult times.
- Was the goal always clear? No, far from it; but the rough direction was always clear. That's why it's so important to have a vision and to always check that you live and act according to your own values. It's always true for me – my intuition is a good compass in my life.
- Is it possible to reach your goals faster? Maybe, but in my life, all events and people always come at exactly the right time. Even the difficult lessons always come at the right time.
- Don't expect gifts from life. Getting opportunities is much more important than getting gifts. Seeing and creating opportunities always and everywhere is even more important.
- Listen to your intuition. Always.

It is May 2023 as I write these lines and I've been living in Zurich for almost exactly eight years. Zurich – that city I fell in love with at first sight.

Let's get back to the rational side again – the facts about the last eight years: My son, my little sunshine, was born in Zurich. I found my absolute professional fulfilment in Zurich with my own company, Elite Experts Conferences.

I guess there are already enough reasons to explain my spontaneous emotional affection for this beautiful city right at the very first sight. And once again, real life shows that everything is possible – with preparation, inspiration, patience, perseverance and love! Just don't call it luck.

3.10   THE SEARCH FOR MEANING: YOU'LL FIND WHAT'S MEANT FOR YOU

It wasn't somehow a special magical and fateful moment where I suddenly understood that if I keep going down the path, I'll get to goals, but those goals won't be mine anymore. No, it really wasn't like Hollywood. It was rather small but steady hints that I perceived internally that I was no longer going my way, that I would only use maybe 70% of my skills and talents in my job in R&D in a prestigious Swiss company and that I would miss more and more something essential.

That essential element was then finally the deciding factor! That essential element was lack of meaning, lack of purpose. I played through the following scenarios – what if I were to develop the super adhesive of all adhesives for the automotive industry? Or deliver the best technical service ever for an OEM customer? Do I really want to 'go down in history' with THAT? Really? Should that be the goal of my life?

No, not at all!

What I have always been good at, saying that now without ego or arrogance, but also without the famous imposter syndrome, is to assess my own abilities very rationally and accurately. I know that I am good with chemistry and I can develop something practical, but my true strengths lie in working with people, not in working with molecules. Even my boss at the time told me once in an evaluation interview that he found it totally fascinating and great that I could walk into any meeting with customers and talk to them in such a way that we would immediately look for solutions professionally at eye level with each other.

I respect myself, and in doing so I show others how they are allowed to treat me. This may sound like an innate talent, but it's actually just the ability to listen to others and always want to understand the other person's side first instead of describing my own situation. Applied empathy and compassion. For my OEM customers I was not only responsible for the global R&D projects, but also for the technical service. Now, do you think that the customer comes to say that everything is fine with the product? Of course not. Customers come only when there are problems in the application. Listening to understand is the first step to de-escalate the situation, and I learned that well.

I stood there with this realization that I would have to correct my path in order to be more myself again. Do you think now that I would immediately think in the direction of entrepreneurship and starting my own company? No, interestingly enough, I approached the matter quite differently at the beginning. At first, I assumed that while I would make a career change to be more in technical marketing, I would look for my purpose outside of my work. I don't know if that was one of my false beliefs from back then, but somehow it didn't make sense to me that I could have both – meaning and work – in one place.

What I knew very clearly was that I want to help people, I want to give something of value back. I have always financially supported numerous charitable aid projects. That's how I was brought up: if you have something, share it with people who have less. Since I belonged to the 'childless top earner' status in Switzerland at that time and I don't have any particularly expensive hobbies, I was able to support even more projects financially. But simply transferring money on a regular basis doesn't give you a sense of meaning. I wanted more.

I actively looked for projects where I could help not only with money, but actively with my time and knowledge. I found projects in Zurich where they help immigrant children learn German, or write job applications or learn computer skills, and much more. The idea of teaching German to someone with a migration background particularly excited me. After all, it was my own path, which I also followed until today, I speak on a native speaker level and without an accent. So, I searched out all contacts, after my work, I wrote super-motivated emails to the project managers at such charity organisations, and I was full of enthusiasm to help. I saw myself soon helping children.

What came back was more than disappointing. Either nothing at all came back or I got a polite answer along the lines of 'your money is always welcome, but we can't use your time and knowledge anywhere at the moment.' Really? So much for volunteering and wanting to help people. I was disappointed and didn't understand the world anymore. I can remember a phone call with my mom when I told her in utter disbelief that apparently no one would appreciate my willingness and desire to help. She just said, '*If your knowledge isn't needed here, it's only because it's much more appreciated and needed somewhere else. You'll find what's meant for you.*'

This is how my search for my path began. What followed was what I briefly outlined at the beginning. I searched for a suitable advanced education based on my interests. This was also very experimental; after all, I have thousands of interests and a natural insatiable curiosity. That was until I discovered this fateful CAS in event management, which combined so many topics that I found exciting: modern marketing, project management, social media, leadership, creativity and more. In the fall of 2017, I enrolled in the program. In January 2018, things were about to start.

Two weeks after enrolment, I knew I was pregnant. Wonderful news; after all, it was PLANNED. I never doubted for a moment that I would be able to do this study in parallel with my work while being pregnant. It was also a brilliant coincidence that this study would have a short break in the summer – that fell exactly on the time with my due date. As if the universe took care of it itself. So off I went. I was pregnant, worked 70%, studied 30% event management and, in order not to get bored, I also took an English-language course to take a C2 exam. That describes quite well what kind of energy level I had.

I was feeling great, and life was beautiful. With all the processes that were going on in my body, my inner being was also changing. My vision became clearer. Yes, I can say that exactly in this very uncertain time, having my son was the best thing that could have happened to me. It made me whole as a personality with all my facets. It made me even stronger. I was also reborn as a human being.

What also became clearer is that I no longer wanted to work in the corporate structure. I saw my colleagues – young fathers and mothers – fighting, really fighting to reduce from 100% to even 80%. Then, there were fights about working from home for even one day per week. Yes, this was 2018 before the pandemic, and today it sounds ridiculous, but at that time the company said this was not possible in the job as R&D project manager.

This rigid inflexibility made anger rise up inside me against the system. No, no one will tell me how to work, I thought to myself. It's not about working less – I'm certainly working 120% at the moment – but it's about working more flexibly. I love my work, and if I am willing and have the ambition to work so much, then it must be something very special – my own company, where I would combine skills, knowledge and purpose.

If anyone thinks it's all easy now, I have to disappoint here as well. The path to success and happiness will not always be easy. It will require dedication and hard work, but the end result will always be worth it.

If someone wants to tell me that you can either have children or make your dreams come true, then I usually say just that my son is almost five years old now, my company almost four years old. It all works if you put your energy into it. No, I don't do giant steps, and I don't do all at once, but I learned to take small, steady steps, every day. Consistency really does win. The whole magic is never in the great plans, but in the steady persistent execution of small steps. No glamour, but hard work.

Now the story should actually end. Like in a fairy tale where everyone lives happily ever after. Yes, Elite Experts Conferences was born in the fall of 2019. The pandemic came in March 2020, and all of a sudden the world stood still. If you organize events and suddenly no events can be held due to the pandemic, you are out of business. Hard reality.

But I felt absolute calm instead of panic. Yes, a 'what we do' – events – was taken away from us. But the 'why' we do something was stronger than ever. My why is to connect people, to create a more human exciting kind of tech marketing, to get people excited about sustainable technologies. All of that hasn't been suspended by the pandemic. All of that is still possible. Yes, differently – namely digitally. So, our Elite Experts Conferences technology podcast was born.

In the meantime, it has become a popular knowledge exchange platform where well-known companies come to us to share the sustainable technological highlights and trends. Every day I feel the meaningfulness of my work, every day I learn something new; and thanks to the pandemic, we have been tested and become stronger. I thank with all my heart these difficulties that have pushed us much more in the digital direction. As a result, we have also made our marketing very digital.

I would also like to pass on to the reader that not everything can or has to be planned down to the last detail in order to turn out well in the end. Did I know or plan from the beginning that I would do large trainings on 'LinkedIn Marketing' or 'Corporate Influencers' or even give coaching to the most exciting leaders of the automotive industry at C-level? No, of course not! This is also a funny and absolutely not planned coincidence.

By surviving through the pandemic, I built such a strong presence on LinkedIn that marketing departments from other companies approached us to ask how we were doing it. In the beginning, I was mostly giving trainings to startups and scaleup companies. But there were more and more requests. So, we were approached by several large tier 1 suppliers, where I now give trainings and coaching. My advantage is that these are almost all people with a technical background, so I speak exactly their language – even when we talk about marketing and communication.

What I totally underestimated is how much joy it gives me to help people with this content, with this knowledge, to see how their lives are transformed for the better through their LinkedIn presence and the right positioning. No, I don't teach them to follow my path. I teach them to recognize their own values and to find their own way. It's not for nothing that I say: I #DERRtobedifferent, I dare to be me.

This magic should also be felt by each individual. Something very special happens, and you attract people differently when you dare to be simply yourself.

Thank you to the universe for this beautiful and extraordinary life. I look forward to everything that is yet to come.

There is not one right path, but there is YOUR path somewhere – it's worth finding it. If I can do it, so can you. Have courage, it gets a little easier with each step.

3.11 EPILOGUE

Here in this chapter, I have written down my personal story, my own career path in STEM. Once again, I would like to point out to the reader that it is one possible path. There are thousands of other paths and other life models. This is the right path for me based on my values and based on my goals. In your case, dear reader, something absolutely different might be right for you. Don't be frightened, upset or distressed if you don't yet know exactly what the right vision is for you in the end. It's also, at least as important, to know what it isn't.

Nevertheless, take small steady steps forward, follow your curiosity, try something new, test in a small format, in small projects, whether something suits you. When you discover strong interests, follow them, invest more time and always listen to yourself, to your own intuition, if it is the right thing to do.

Also work early on your supportive environment, your circle of friends and business partners, that takes your enthusiasm for something new seriously and is rooting for you to succeed. The more like-minded people around you, the better. The more people around you that are already going the way you want to go or are already living the life you dream of, the better it is, and the easier you will succeed. Forget about envy; it's just a waste of energy. Instead, think of it as inspiration and motivation.

That's why it's also important to keep giving knowledge back to the universe. There is always someone who is one or more steps behind you – so help others with your experience, your wisdom. This can often have an immense effect. We know this ourselves – sometimes a sentence at the right time in the right place can trigger strong insight in us. Well, we can and should give the same back to others. *At the end of your life it is not what you own or what you have achieved that counts, but how many lives you have influenced for the better.* May it be many lives.

ABOUT THE AUTHOR

Dr. Ludmilla Derr is Founder and Managing Director of Elite Experts Conferences and the host of the Elite Experts Conferences Technology Podcast. Ludmilla has a technical background with a diploma degree in chemistry, a PhD in materials sciences and eight years of experience in applied research. Ludmilla also brings very practical experience from the automotive world. After her PhD, she worked in Zurich at Sika Technology R&D headquarters and was responsible for R&D projects worldwide for the customers BMW Group and Fiat Chrysler Group in adhesive technologies in automotive trim applications.

4 Paving the Way Forward for Other Female Arab Engineers

Eman Martin-Vignerte
Bosch UK & Ireland, United Kingdom

I am a proud Arab woman with a background in electrical engineering working in the automotive industry, which is a traditionally male-dominated field. Glass ceilings are the limits placed upon us by society. They are the societal "rules" by which we have lived for centuries and are treated such that they should not be broken. However, I believe that we should not place limits on members of our society; otherwise, we cannot evolve and move forward together.

4.1 FROM QATAR TO GERMANY – PURSUING MY EDUCATION AND DREAMS

My journey began when I was in high school in Qatar and discovered my passion for math and science. In fact, I was ranked as the second-best performing student in completing science in my year. I knew that I wanted to pursue a career in engineering and started looking for opportunities to learn more about the field. I was fortunate that my father was very supportive and allowed me to study in a foreign country. His support and approval have paved the way forward to allow me to return the favour to other young Arab women who want to pursue careers in engineering.

After completing my secondary education, I had the opportunity to study electrical engineering in Paderborn, Germany. It was an exciting new adventure outside of my home country, where I could be immersed in a different culture and learn more about engineering. I remember distinctly that for the entire semester, we had only 5 females out of 400 students. This made me question whether I had chosen the right field of study: Was this the right decision? It even made me question whether I would be accepted by my community for choosing such a predominantly male field.

However, I did not let these questions stand in my way. It was reassuring to lead the way forward for other females and to encourage each other to continue along this path. Although it was a luxury to have female colleagues, it made the journey more tolerable. I worked hard and graduated at the top of my class with a degree in electrical engineering. I continued on to study medical engineering.

At that point, I wasn't sure whether I should go back to my home country or stay in Europe. Would it be easier to get work experience in Europe, or should I try to break into the industry as a female in my country? It was clear either way that the

DOI: 10.1201/9781032679518-5

journey would not be easy. It was clear to me then that women need to work harder than males in these fields. While it has improved a bit now, there was still a lot of reluctance during my time. One of the main concerns was whether female engineers were able to deliver the job with the possibility of potentially starting a family soon.

4.2 BEING ONE OF A KIND IN THE AUTOMOTIVE WORLD

My first job was with a Tier One supplier in the automotive industry. I was excited to be working in an industry that is constantly evolving and pushing the boundaries of technology. I was responsible for software and hardware development for engines in cars. The job was incredibly interesting and innovative. However, I was one of a kind, as there were not many of us there.

As an Arab woman in a male-dominated field, I have faced many challenges. I was often the only woman in the room and had to work twice as hard to prove myself. I also had to overcome cultural stereotypes and biases that made it difficult for me to be taken seriously.

I refused to let these challenges hold me back. I was determined to succeed and make a positive impact in the industry. I worked hard to build my skills and knowledge. Over time, I gained the respect and trust of my colleagues. The lesson learnt was to continue pushing boundaries and breaking the glass ceiling as this is the way to allow more women to enter these fields.

4.3 TIPS ON BREAKING THOSE GLASS CEILINGS

Breaking the glass ceiling can be a difficult challenge, but it is not impossible. Here are some learnings which helped me to break the glass ceiling:

1. *I learned that I needed to build my skills and knowledge*: Make sure that you have the necessary skills and knowledge to excel in the industry. This means staying up-to-date with the latest trends and developments, and continually improving your skills through training and education. Improving your skills is an investment in your career and will help you to gain more knowledge about your field.
2. *It is important to have a network and build relationships*: Building strong relationships and networks are crucial for breaking the glass ceiling. Attending industry events, connecting with mentors, and building relationships with colleagues and peers are just a few ways that you can become better networked. Connecting with your mentors can also lead you to exciting new opportunities as they may be aware of things occurring within the industry and can recommend you to be part of moving the industry ahead.
3. *Find a mentor*: A mentor can provide valuable guidance and support as you navigate your career. My advice is to look for someone who has achieved success in the industry and can offer advice and support. Mentors can also be mentors without having an official title. You can follow someone's career and look up to them without joining a mentorship program. Mentors are usually advocates who help others in the industry. Reach out to those who

inspire you and ask for advice; their answers may help to lead you on a journey of self-discovery.

4. *Speak up*: Don't be afraid to speak up and share your ideas and opinions. Always know that your perspective is valuable, and speaking up can help gain visibility and credibility. Everyone has a unique perspective, but if you don't speak up, then you are not sharing your perspective with the team. By sharing your perspective, you are essentially working together to help improve or solve the challenge ahead. While this builds your teamwork skills, it can also help to boost your confidence and allow team members to recognize your abilities.

5. *Seek out opportunities*: Look for opportunities to take on new challenges and responsibilities, even if they are outside of your comfort zone. This has helped me to gain new skills and experiences, and to demonstrate my value to the organization. Sometimes these new opportunities can lead to networking with the right people and getting you the visibility for your skills and talents to be recognized.

6. *Be persistent*: Breaking the glass ceiling can be a long and challenging process. It is important to stay persistent, stay focused on the goals, and continue to work hard, building the skills and network over time. It does not happen in the blink of an eye, and there is a lot of work involved in breaking those glass ceilings. If it were an easy task, then everyone would do it.

Remember, breaking the glass ceiling is not just about achieving success for yourself, but also about paving the way for others who will follow in our footsteps. By breaking the glass ceiling, we can inspire and empower others to achieve their own goals and make a positive impact in their industry.

I am proud to say that I have achieved a lot in my career thus far. I have worked on some of the most cutting-edge projects in the industry, and my contributions have been recognized by my company and my peers. I have also become a role model for other women in the industry, showing them that it is possible to succeed and thrive in this field. I evolved in different industries, e.g. automotive, healthcare, and energy, and now I oversee the political affairs activities and use my knowledge to define and develop the future policy frameworks and innovation future trends and solutions.

Looking ahead, I am excited about what the future holds. I believe that the industry has the potential to make a real difference in the world, and I am proud to be a part of that. As a female leader, I hope to bring a unique perspective to the industry and help shape its future in a positive way.

I want to encourage other women, from all different communities, to pursue careers in engineering and the automotive industry. It's not always easy, but it is rewarding, and the impact you can make is truly incredible.

ABOUT THE AUTHOR

Eman Martin Vignerte is the head of political affairs and government relations for Bosch UK & Ireland. She holds B.S and M.S. degrees in electrical and medical engineering from the University of Paderborn, Ulm, Germany. She was responsible for business development for telehealth in the UK and Ireland. She has more than six years of experience in software and hardware engineering in the automotive industry. In 2004, she was the project manager for electronic pumps, Hyundai customer, and moved for a year to Korea. She was involved in developing the ceramic Medtronic control units for vehicle engines at Bosch GmbH. She is a board member at BBF (Buckingham Business First). She is the Chair CBI (Confederation of British Industry) Southeast, board member of local enterprise partnership Buckinghamshire, board member at Success Essex Partnership, and an advisor at AUTOEUROPE. She also sits on the advisory board for UK5G. She is a member at OCAVIA – The Oxford-Cambridge ARC Virtual Institute for Aviation, and is an advisor for WE-Transform, Workforce Europe transformation agenda for transport automation, EU Commission.

5 Fueling the Future of Women in Industry

Stephanie Hajducek
This One's for the Gals, USA

The vibration from my cell phone woke me up way too early on a Saturday morning back in 2006. As I reached for my nightstand, my first instinct was to decline the call so I could roll over and go back to sleep, but I must have hit the wrong button because the voice of my friend Matt, a guy I'd known since elementary school, came blaring through the speaker. "Stephanie are you dressed to go to Bechtel's job fair?" he asked. I remember putting my hand on my head as I was about to break the news to him that I decided I wasn't going. I'd never been to the Galleria area in Houston, much less stepped foot inside the building of one of the world's largest engineering companies. I was a small-town girl from Van Vleck, Texas. I grew up in the country and graduated in a class of 62 people. I had no business going to this type of job fair because more than likely I'd have zero chance of getting hired. Or so I thought!

5.1 HARNESSING THE WILLINGNESS TO LEARN

A few months earlier, Matt had encouraged me to take a six-week AutoCAD drafting and design program. At that point in my life, I was a single mom with two little boys, twins, and I knew that if I was going to provide a good life for them, I needed a career and not just a job. I decided to take the AutoCAD course and found not only that I enjoyed it, but I was pretty good at it, too. Now, it was time to get out there and look for a job, and in all honesty, I was terrified. Coming out of high school, I didn't have any idea what I wanted to do for a career, and although my parents talked a lot about college, I didn't do a whole lot of listening.

All I knew was that I was an adult and ready to get out into the world; everything else I'd just figure out. Needless to say, that's the worst plan ever. I ended up getting married young, having kids and then getting divorced. I was a single mom with two kids, no post-secondary education or training and a very feeble resume with no clue as to what I was going to do. But I knew I had to do something! I had two little guys depending on me and I was not going to let them down. Ultimately, I decided to get up, get dressed and head to the job fair. Matt said all I had to do was walk in there and tell them I wanted an entry-level job. That sounded easy enough.

So, there I go, off to Bechtel's main office in Houston on Post Oak Boulevard. Now remember, I'd never been to this area of Houston and pretty much had no clue where I was going. I ended up parking in the wrong parking garage and was immediately met by a not-so-cheery security guard who proceeded to tell me that I couldn't park in her lot for the job fair and that I needed to go find somewhere else to park.

DOI: 10.1201/9781032679518-6

At that point I almost got in my car and drove back home. I didn't even know the right place to park; how on earth is this venture going to turn out positive?

But I didn't go home. I found the right place to park and walked inside Bechtel's 17-story building with my very feeble resume, my six-week AutoCAD certificate in one hand and a willingness to learn in the other. I told them that I was there for an entry-level job, if they had something available. A couple of weeks later Bechtel called and offered me a job, and my life was forever changed. It changed everything about what I thought it was to be an adult, to have a career and to be a woman in the workforce. Within one year I bought my boys their very own house, and about a year after that I met the most amazing man who would very quickly become my husband. We'd later complete our family tree by adding a daughter to our crazy, beautiful life.

5.2 DREAMS DON'T WORK UNLESS WE DO

I'd go on to work for Bechtel for 12 years and progressed far beyond that entry-level job. I worked hard to differentiate myself from everyone else and took on tasks others didn't want to do. I came in early and stayed late, never once complaining because I knew that dreams don't work unless we do. I had an amazing career with Bechtel and learned so much about myself and just how much I was capable of.

Since my time at Bechtel, I've held great roles with several other awesome companies such as ChevronPhillips, Puffer, Citgo, Steel Dynamics and The Chemours Company, and now I'm in an amazing role as a Territory Sales Manager for Samson Group.

In December 2018 at the age of 38, I walked across the stage at the University of Southern Mississippi to accept my bachelor's degree in industrial engineering technology. I can still remember hearing my husband, my kids and my mom and dad cheering in the stands. Although that was such an amazing accomplishment, I keep thinking of how much time it took me to find my place in the workforce. I feel that if I had had a little better direction coming out of high school, I wouldn't have spent so much time trying to figure things out.

In January 2022 I decided to start a nonprofit organization called This One's for the Gals (TOFTG). Our mission is to help girls with career exploration and workforce development in the construction, energy and manufacturing industries. Specifically, we want to assist girls from low-income households and disadvantaged backgrounds into stable, high-paying careers in an effort to conquer generational poverty. We highlight occupations that girls didn't consider previously, didn't know existed or didn't think were for them. I like to say that we talk to girls about careers that not a lot of people talk to girls about!

I've never considered myself an entrepreneur, a fundraiser or a nonprofit launcher by any means. It just sort of happened. I had a small idea to help get some high school girls to a conference where they could learn about different careers. I started out thinking that I could get 10–12 girls to this conference, but by the time I was done, I was able to get over 200 girls to the conference. This was made possible through a collaborative effort between our local school districts and industry partners who helped provide funding. Many of the schools made it an overnight trip where they stayed in a hotel.

The night before the conference as the students were hanging out in the pool at the hotel, one of the counselors came up to me and thanked me for helping get the initiative started. She then said something that I will never, ever forget. She told me that I may not realize it, but the majority of their girls had probably never stayed in a hotel room before, with many probably never having traveled before out of their hometown. I had not thought of that, but that is the dynamic in many parts of our area, especially in our small rural communities.

I like to say that that's the night that TOFTG was truly born, because it was no longer about a conference. It's about creating awareness and opportunity. It's about getting our girls to stop being okay with just being okay and just getting by. It's about getting our girls to want more out of life; to not just talk to them about career opportunities, but to show them how to go out there and get the training and education to prepare them to enter the workforce.

5.3   CREATING AWARENESS FOR THE FUTURE GENERATION

My passion for TOFTG is fueled by my being that girl that I so desperately want to help take charge of her future and reach for the stars when it comes to her career. I had a simple yet amazing childhood. My family lived on thousands of acres of land, but that land didn't belong to us. It belonged to the rich man down the road who my grandpa worked for and who also became his best friend. He worked the land, and our family was able to live there in exchange. My grandparents didn't graduate from high school and spoke very little English.

My parents were first-generation high school graduates. My dad worked for an auto parts store, and my mom was a cake decorator for our local grocery store. They worked hard and provided a great life for me and my older sister. (I now have a younger sister who would come along when I was 21, but that's a whole other awesome story in itself!!) I love everything about my childhood and how I was blessed to grow up. Times were simpler back then, and what got us by back then won't get us by now. Not only that, but there is so much opportunity for the next generation and creating awareness of different career paths is crucial to their success and ours!

I read something on Facebook a while back that said something along the lines of this:

> We don't think we are better than anyone. We aren't trying to compete with anyone. We are simple people from small towns and humble homes trying to win the battle that our parents couldn't win. The battle against generational poverty.

Don't get me wrong. The young women that I am referring to aren't living a life with a shortage of love and care. Their parents are working hard to provide for them. Their grandparents are the Martin and Nicanora Mireles's of the world, my grandparents, and their parents are the Joe and Irene Mireles's of the world, my parents. But just like my grandparents and parents didn't quite know how to help guide me into a career, they may not have that help, either, and that's okay. That's what TOFTG is here for! We want them to know that they can do hard things, and we are here to show them how!

ABOUT THE AUTHOR

Stephanie Hajducek, Territory Sales Manager for Samson Group, is the Founder and Chief Visionary Officer of This One's for the Gals, a Texas-based 501(c)(3) nonprofit organization that helps female students in grades K-12 with career exploration and workforce development in the construction, energy and manufacturing industries. At the age of 38, Stephanie walked across the stage at the University of Southern Mississippi to accept her bachelor's degree in industrial engineering technology. While she realizes that was an amazing accomplishment, she can't help but think of the number of years she spent spinning her wheels trying to find her place in the workforce. Her goal now is not only to help girls with career exploration but to introduce them to occupations that they may not have considered previously, didn't know existed or didn't think were for them. Stephanie hopes that by sharing her experiences and lessons learned she can help girls step out of their comfort zone and step into a world of endless possibilities in industry!

6 Building the Next Generation of STEM Heroes

Marcella Ceva
WE Ventures – Microsoft, Brazil

As part of an exercise at a networking dinner with a group of women in finance back in 2017, we went around and spoke about our mothers' careers. Coincidently or not, we realized that all 30 of us had strong, hardworking mothers. This was really new information for me, as I wasn't a mom yet, but it stuck with me for years. Was this just a big coincidence, or do we play such a big part in shaping our children's futures?

Melinda Gates mentions in her book that schools with female math teachers statistically graduate more young girls into STEM careers.[1] Today, as a mother of two, this is what I try to bring into my home on a daily basis.

6.1 THE JOURNEY FROM BEING AN M&A LAWYER TO FINANCE TO INVESTING IN WOMEN IN TECH

Careers in finance are hard. There's a lot of pressure and responsibility, long hours, lots of travel and constant education. To be completely honest, I am not technically a STEM graduate. I studied international relations and then went to law school, even though I always enjoyed math throughout my school life and always took AP classes, pre-calc and calc and all that. On a side note, I actually ended up in law school because I wanted to be a diplomat.

But starting out with an internship in corporate law, I loved it so much that I just kept going. Corporate law steered me into the legal aspects of M&A transactions – mergers and acquisitions – and as an M&A lawyer, I decided to transition into the financial advisory of such transactions. So, I took up corporate finance, financial modeling and valuation classes and landed a job in investment banking at a firm that really valued my legal background and invested in my interest for a financial education.

Fourteen years (and some long sleepless nights) later, I am now proudly head of the first venture capital fund in Latin America to invest exclusively in women in tech. This is at the same time a great honor and a great responsibility, since only 2% of global venture capital resources go to female-led startups and the gender gap we set out to close is immense.[2]

DOI: 10.1201/9781032679518-7

(As I sat down to write this, I concluded I am definitely in STEM and didn't need to make this disclaimer, but the lawyer in me couldn't help it.)

As a woman head of an investment fund, I am part of a select group of 8% of women worldwide in decision-making positions in financial institutions.[3] If we look at the boards of directors of Brazilian companies, we will find only 7% of board members are women. Sadly, if we carve out all the women who inherited their businesses from their fathers, we will be left with 3% of board members in Brazil being women.[4] In spite of these statistics, I never felt anything less than part of the teams of men I've worked with.

I think this is definitely due to my upbringing and the values we shared at home, since I come from a family of five siblings that were brought up equally by working parents who motivated us to study hard and work just as hard as they did. No gender-based distinctions whatsoever. Growing up in my house in the 1980s, girls could play soccer and skateboard. No questions asked.

6.2 SWITCHING ROLES FROM A HIGH PERFORMER TO GOING ON A SABBATICAL

Naturally, I bring a lot of what I experienced into raising my kids. I've always been a high-performer. I've always enjoyed studying and working. In school, I took on AP classes and honors courses. I loved the challenge and loved the satisfaction of the awards and praise even more. Not that my parents were huge on praise, because at home we lived by the maximum that as kids, school was our only job. So, no outstanding performance was rewarded simply because it was nothing short of our only obligation.

Used to being a high performer in school, I definitely carried those practices into my professional life. Reviewing everything I did multiple times and proofreading and rewriting and correcting and tweaking – to me, no job was done until it was really finished up to my highest standards. Working without kids was fantastic and pleasurable. Sure, I had long hours, but work was my life, and I loved it dearly. Everything else was accessory to that pillar that was work. That was until my first child was born.

I was actually on a sabbatical when my first child was born. Quitting my job in finance was a really tough decision to begin with, because I loved it so much and couldn't see myself doing anything else. But a sabbatical year was a longtime dream, and I thought it would also be the perfect time to try to get pregnant, since for as much as I loved it, I really did not believe the investment banking lifestyle was compatible with the vision I had for motherhood.

My sabbatical was amazing, by the way. I studied and volunteered and traveled and crossed off multiple things from my bucket list – and from my simple luxuries list – like reading books while taking long baths. Things unimaginable for a person working in finance. And that's actually the reason sabbatical years are so common in our market. Some of us want to travel the world and see migrating butterflies. Some of us just want to soak in a tub.

6.3   BEING A PARENT ISN'T INCOMPATIBLE WITH WORK AS WE KNOW IT

I always knew I wanted to be a working mother. And by this I do not mean that being a stay-at-home mom is easy, because now I know that being a mother is the toughest full-time job on the planet. But I simply wanted it because work has always been such an integral part of who I am that I can't really ever imagine myself without it. When I became pregnant, around 10 months into my sabbatical, I started slowly putting together a list of jobs I would like to have as a working mom.

This ranged from anything between going to medical school to become a doctor, joining the legal or M&A teams at a multinational company or simply going back to investment banking (because it paid so well and was so seductive and safe and totally in my comfort zone, and I had multiple offers). Do note that nowhere in my list was leading a venture capital fund, even though that's where I – thankfully – ended up.

I researched great places to work and looked into firms that offered the most amazing benefits for working mothers. I talked to people in many different industries and learned a lot about a whole new world of corporate policies that exist to make going back to work easier for new moms. Did you know approximately 43% of women give up work at some point in their careers because of motherhood?[5] That is how the study phrased it, but in my opinion these women do not give up work because of motherhood. They give up working in places and industries that are outdated and lagging behind in terms of maternity and paternity benefits. It's not being a parent that is incompatible with work. It's some kinds of work that need to be adjusted to the reality of life.

6.4   ENTERING THE VENTURE CAPITAL WORLD

My son was around three months old when I was invited to the launch event of a venture capital fund focused on women in technology at Microsoft offices in São Paulo, Brazil. It was the first time I left the house for a work-related event, since I was still enjoying my full-time sabbatical/motherhood/exclusive breastfeeding experience.

I planned a three-hour interval between feedings, and off I went. I was excited to do something other than baby-related activities, but also nervous to leave him for the first time. Even though I had to leave a few minutes before it was over because of my breastfeeding schedule, I loved every second of the presentation and was so inspired by this initiative that I even mentioned it to my friends in the women in finance group.

Nothing much to that. Life went on as normal. Breastfeed, pump, repeat. The next day, I got a call from a former colleague in banking, who was now the fund manager for this Microsoft fund. He thanked me for being there, appreciated the effort of leaving the baby at home, and casually told me he was looking for a woman to lead this fund. Just as casually, I told him I had many names to suggest.

Little did I know that he was actually thinking of me. This casual conversation turned into a really unexpected job offer. And I panicked. How on Earth would I get back to work and cut short my dreamy full-time motherhood experience? Well, it turns out that after a couple of months and lots of old-fashioned courtship, I joined the team, and I'm so grateful to have done so.

6.5 MAKING TOUGH DECISIONS

Throughout my career, I've had to make innumerable tough decisions at work. And especially in finance, such decisions impact not only human beings but also people's and companies' assets, right? Sometimes hundreds of millions in assets. But believe me when I say this: nothing compares to how hard it is to leave your child at home and go to the office. To give him a bottle when you wanted to breastfeed. To miss the first time he eats a piece of fruit. To miss many baths and bedtime stories. To board a plane and leave your child for a work trip and, consequently, to miss all those baths and meals and playtimes and snuggles.

Being a mother in finance is hard. Being a working mother is hard. Well, that's because being a mother is hard. Period. And I strongly believe that we need to be vocal about this. Nobody ever told me how hard it would be. Women show up daily at the office for 9 a.m. meetings in their neatly pressed outfits and perfect hair and makeup, and nobody talks about what goes on in a house between 6 a.m. and 8 a.m. The complete chaos of getting other human beings ready for school and feeding them and packing school bags and finding uniforms and all that. Not to mention that those same humans are likely to wake up in the middle of the night multiple times a week.

I was definitely not prepared for this. Not only for the tough decisions, but also for everything I had to give up. As a high performer in all areas of life – and please note I am trying to stick to "high-performer" instead of "perfectionist" – since becoming a mother I have learned that there is no such thing as perfection. Anything you do as a mother is perfect. Mothers are so privileged to be trusted by God – or the universe or whatever it is that you believe in – with the well-being and upbringing of another human being. We try our best. And I honestly think that anything we deliver with love is pure perfection.

6.6 MOTHERS ARE ALWAYS TIRED

But yes, as a high performer, I was used to being effortlessly good at my job. And I am by no means arrogant. All I mean is that all my energy was channeled into work, so it was easy to be good at it because I had all this time to invest in it. After my first child was born and I started working, I had a really rough first few months. It was so frustrating because I wanted to work 12 hours a day to deliver what I was used to delivering.

But I also wanted to spend as much time as possible with my baby. And I wanted to exercise and have time for myself and have a social life, go out on dates with my husband and see our friends. And obviously I had a new household to take care of, a family home, with meals and groceries, not the takeout festival of the recently married couple we were so used to. To top it off, I was exhausted from all the sleepless nights. And nothing sours my mood like lack of sleep. In sum, I was a grumpy mess. For months.

Back then, I read something that really touched me about a mother in a similar situation. She used an expression to describe herself that was so visual to me: she was spreading herself too thin. That was exactly how I felt. I was like the stretchy mother in *The Incredibles*, only I was stretching way beyond what my already superhumanly stretchy body could handle. I looked around and all mothers seemed fine to me. I asked them if they weren't tired, and all I got every time was: "mothers are always tired". Wow. Was that really it? I felt so guilty all the time. Guilty for not spending enough time with my child, guilty for not working 12 hours a day, guilty for being tired when I should be as happy and perky as all other mothers seemed to be.

It took me a good full year of therapy to learn and accept that my life would never be the same. I have a new life now, and it is amazing and fulfilling, and I'm the happiest I've ever been. But I am not the same as I was four years ago. I sometimes drop the ball. Sometimes there's no dinner. Sometimes I miss important deadlines because a child is sick. Sometimes I forget to check in for a flight and end up in seat number 43K. I drove my son to school the other day only to find out I'd forgotten his shoes. I take a couple of days to reply to most of my email. Some texts go unanswered for weeks. For a former perfectionist, accepting this is hard. But you know what? It's okay.

6.7   CLOSING THE GENDER GAP

I have two healthy wonderful kids and I make room in my schedule to spend as much time with them as possible. I have a dream job in finance with a strong sense of purpose that allows me to be part of the change we so desperately need. I have a loving family, and we have the social life we are able to at this point in life. I exercise as much as I can. I eat as healthy as I can. I really do the best I can.

As a mother working in a STEM career filled with responsibility and timelines, I had to learn to be more gentle on myself. In all areas. Every time I find myself being critical and too hard on myself, I have to take a step back and remember to actively be more gentle. It's still not natural or intuitive. To be completely honest, it is a daily exercise. But it's worth it.

I consider myself privileged to have such a fulfilling job through which I touch the lives of so many women. This is the one thing that motivates me to work at this moment in which my kids are so dependent on me. I am being part of the change. I am working so that the world will have a more sustainable relationship with finance. I am contributing to an increase of $2 trillion to $6 trillion in global GDP by including women in the economy.[6] I am connecting so many women with large corporations. I am closing the gender gap in finance.

Young women entering STEM careers ask me all the time how this is all possible. I consider it to be my mission to be really vocal about how hard it is – all the choices and dilemmas and daily struggles. If women aren't vocal about our own cause, then who will be?

6.7.1   BUILDING THE NEXT GENERATION FOR A BETTER SOCIETY

We can start small, in our own homes. Every relationship we build with our partners and children help shape a more equitable society in the future. The values we pass on do not depend solely on what we say, but mostly on our actions. Children are fantastic observers that will replicate what they see at home between their parents.

Fathers who share responsibilities in the house, who actively participate in raising their children, will create the next generation of naturally engaged and present fathers. Mothers who are vocal about current and historical inequalities in the household will raise children that are aware of such problems and will therefore behave differently. These kids make up the next generation of activists that will take over our mission to create a more equitable society.

6.7.2   BEING MORE VOCAL IN THE WORKPLACE

A second step for women is being vocal at work. We've had enough missed school events or pediatricians' appointments, especially in finance. The more vocal we are at work, the more pressure companies and financial institutions will feel to create a more empathetic environment for mothers. Most policies are financially irrelevant for companies and would make a huge difference for working mothers. Especially in the post-pandemic world, tools such as working from home should be more broadly accepted for parents.

Extended maternity leave and mandatory paternity leave are also essential. In developing countries such as Brazil, the few companies that do offer paternity leave have almost no candidates for it because men are so afraid to receive a pay cut or negative feedback. Other benefits may come in diverse forms, be it lactation rooms, childcare aid, maternity support apps, community events, bring-your-child-to-work days – there's so much that can be done to reach out to working mothers and make them feel cared for. And companies will subscribe to this only if we ask for it. As long as we continue to hide our maternity journey to make us seem tougher at work, no change will come.

6.7.3   ENGAGING WITH ADVOCACY AND PUBLIC POLICYMAKING

A last – but not final – step for women toward gender equality is engaging with advocacy and public policymaking. I understand we are tired and overworked with our multiple journeys, but we have all the tools to be the catalysts of this change. The more women engage at home and at work, the more companies will also engage and consequently the more constituents will engage in our behalf. Today, women worldwide have around 75% of the rights that men have. This seems bad, but we've come a long way from the 50% we had in 1970.[7]

Only recently did women gain the rights to own property and to drive in certain areas of the globe. Changes in policy are fundamental to coerce those who do not voluntarily adhere to gender equality to forcibly do so. Just as we have observed so much positive change in European countries' environmental policies in the last 10 years, the more we include gender equality in this agenda, the closer we will be to a more equitable society.

If I could leave a single message to young women in STEM, it would be to have faith in yourselves. Obviously, study hard and be three times as prepared as any man, because the world is filled with conscious and unconscious bias. But a prepared woman will be stopped at nothing. STEM careers are hard, but we belong there as much as we belong in any other industry. We belong in financial institutions, we belong in boards of directors, we belong in executive positions. A prepared woman will be stopped at nothing. To us, anything is possible.

ABOUT THE AUTHOR

Marcella Ceva is the Chief Investment Officer for Microsoft's WE Ventures venture capital fund, with over 14 years of experience in investment banking, M&A and fund raising in global firms such as Evercore Partners and Squire, Sanders & Dempsey LLP. She began her career in corporate law, has a law degree from Universidade Federal do Estado do Rio de Janeiro, studied international relations at Michigan State University, specialized in human rights and women's health at Stanford, and holds a CFA certificate in ESG investing. Marcella is among LAVCA's top women investing in Latin America, Revista Exame's 100 Women in Innovation, marathon runner, active volunteer and board member for NGOs, and a proud mother of two.

NOTES

1 Melinda French Gates, The Moment of Lift: How Empowering Women Changes the World, 2019.
2 BCG & MassChallenge Report, 2019 | Pitchbook & All Raise: All In Report, 2019.
3 World Economic Forum, The Global Gender Gap Report, 2017.
4 Luiza Trajano and Karla Mamona, Revista Exame, 2017.
5 Sheryl Sandberg, Lean In: Women, Work and the Will to Lead, 2013.
6 World Bank, Press Release 2020/125/GENDER, 2020.
7 World Bank, Press Release 2020/125/GENDER, 2020.

Section II

Non-Traditional Paths

7 STEM with Mr N
STEM Communication, Awards, and the Power of STEM Advocates

Stuart Naismith
North Lanarkshire Council, United Kingdom

7.1 THE EVOLUTION OF TECHNOLOGY

The world of technology is constantly shifting, changing and improving, and at a rapid pace. I'm old enough to remember not having internet in the house, and then using dial-up internet, before getting broadband and Wi-Fi when I was in my late teens. In 2022, less than 20 years after I got broadband, a teacher joined my school who had never known life *without* broadband.

I remember watching *Star Wars* as a young child (when they were remastered – I'm not *that* old) and thinking how cool the droids were. Visions of a far-off future, if we were ever to get there, filled my head. And then recently I've been amazed to see robots from Boston Dynamics doing parkour, including flips!

Only 20 years after buying my first mobile phone, a cumbersome device with very little memory and no functions beyond calling and texting, most primary children I have taught have a smartphone. They literally have small, powerful computers with the world's information in their pockets and at their fingertips. What a change from 20 years ago!

The list of such technological advancements in my lifetime could go on for a while: self-serve checkouts; game consoles; virtual reality headsets; digital platforms where you can communicate your message with people anywhere in the world; the list is virtually endless and will continue to expand.

Like it or not, this is the world we are living in, and our children and young people are growing up in. There are jobs that exist today that did not exist when I was a child, such as an AI prompter. AI was a far stretch of the imagination during my childhood. There is an increasing skills gap between what industry requires and skills of our young people coming through the education system. There is a lack of diversity of genders and ethnicities within certain STEM disciplines. Statistics recently reported by WISE (Women into Science and Engineering) showed that just over a quarter of the STEM workforce, by the end of 2021, were women. This figure was even lower for the engineering and technology sectors specifically, where the numbers dropped to 12% and 21% respectively.

DOI: 10.1201/9781032679518-9

This is why it is essential for our children and young people to be exposed to STEM education and an awareness of the diversity of careers, genders, ethnicities, backgrounds and routes into STEM, and why they need advocates who will talk about their experiences. My work as a science communicator across social media, *STEM with Mr N*, aims to tackle these issues.

7.2   SWITCHING GEARS FROM FINANCE TO THE CLASSROOM

Let me get one thing out of the way right from the start: before becoming a teacher, I worked for a car finance broker and had no background in STEM. So, how did I go from a salesman of car finance to an award-winning primary science teacher and advocate for STEM education and careers?

Throughout high school I was a creative person, with my main focus being on music. I played the tuba in wind bands, brass ensembles and an orchestra, and I played the bass guitar in a swing band and a couple of rock bands. I had close friends and bandmates who were also great at science, but that just wasn't where my interests lay.

Following high school, I attended university to study commercial music, which focused on the business elements of the music industry, such as events management, record contracts and copyright law, with some performance thrown in. This allowed me to study my passion while I was young, while also obtaining transferable skills for moving into the wider world of work, as I was pretty sure I was not going to directly pursue a career in the music industry.

Fast forward a few years and I was working as a successful salesman/account manager for a vehicle leasing broker, assisting clients and dealers to secure finance for their new company cars. This job allowed me to buy my first flat, enjoy nice holidays and drive a nice company car. Outside of work, one of my main passions was cycling, which I was able to combine with the charitable ethos of the company I worked for to be able to raise thousands of pounds for various charities.

Then one night my life took a dramatic change of direction.

As I was cycling home from work – a regular commute I undertook at least three times per week, covering at least 100 miles on the road – a careless driver entered a roundabout and smashed into the side of my bike, tossing me in the air before crashing to the ground. I was lucky to have escaped without major injuries, though I have been left with permanent damage from this encounter.

As I sat at home, crutches by my side, replaying the events of that night over and over in my head, I asked myself, *"If I died that night, would I be happy with what I was doing with my life?"* The answer I gave myself was no, I would not be happy. At that moment, I decided to quit my job and undertake the post-graduate teaching diploma (PGDE) to become a primary school teacher.

7.3   THE POWER OF STEM ADVOCATES

When I undertook the PGDE, I was privileged to receive instruction in our Primary Science class by an incredible woman in STEM, Dr Margaret Ritchie. Margaret was passionate about her subject, and I found her attitude and enthusiasm to be infectious. She made me excited to attend that class, to think differently about things around me and how topics could be approached. One day I happened to ask Margaret a seemingly innocuous question.

Little did I know that asking this question would change the direction of my life for the second time in as many years.

Margaret loved my curiosity, so she invited me to attend a course at the University of Edinburgh all about astrobiology, and how the search for life in space could be brought into a primary school classroom. Then, I was invited by the whirlwind which was Margaret to work with other researchers at the University of Edinburgh to create resources for primary schools linked to their research on various uses of solar energy. One thing led to another, and the next thing I knew, I was presenting at conferences, an astrobiology summer camp and STEM Summer Academies, all before starting my first teaching position.

Margaret's enthusiasm had lit a fire of interest within me, and I decided to focus on developing my STEM education skills within a system where most primary teachers do not have a background in STEM and often lack confidence in the delivery of this part of the curriculum.

I decided to carve out this niche for myself within the competitive job market of teaching, to stand out from the crowd and to ignite the same fire within my pupils that Margaret had ignited within me. I found ways to weave STEM into most of the other areas of the curriculum, read the pupils STEM-themed novels, undertook training with the European Space Agency, ran an astrobiology after-school club, developed a Classifying Stars after-school club in partnership with researchers at St Andrews University and helped a fellow school write their STEM planners.

Not bad for someone with no science background!

That was always Margaret's aim with me. She admitted that I was a bit of a project for her, to show that someone with no science background, but with passion and enthusiasm, can lead on STEM education and inspire the next generation. It is precisely because she believed in me, gave me a push and ignited an enthusiasm and passion within me that I've achieved what I have so far, and I am privileged to be able to pass this on to children and young people on a daily basis, either in my classroom, my school, or via YouTube and social media.

Sadly, Dr Margaret Ritchie passed away in 2023. Margaret had been my lecturer, mentor and friend and, as you can undoubtedly tell, she had a huge influence on my career trajectory. I still think of Margaret often, whenever I am teaching or achieving in STEM, and I owe a lot to her friendship and guidance. I hope this section can serve as a fitting tribute to Margaret's legacy of encouraging and supporting others to engage with STEM.

7.4   DEBUNKING STEM MYTHS

As it turned out, that was just the start of my STEM journey. However, before I go any further, let's clear up some myths about STEM.

1. *STEM education is simply about teaching children about science, technology, engineering and mathematics.*
   It is about allowing pupils to engage their brains in new ways of thinking, engaging with problem solving and drawing on creativity to develop

solutions. It is about teaching a process of planning, developing, testing, analysing and revising processes to continue to develop their skills and knowledge. Running through all of the above should be engagement with the real-world applications of STEM skills and the careers associated with STEM. You may also see this acronym with the inclusion of A – STEAM – where Arts have been included. This acronym is bringing more connections to different curricular areas, which brings me to my next myth.

2. *STEM does come at the expense of other subjects.*

STEM is cross-curricular. For example, pupils should be encouraged to write up their plans, methods, results and conclusions. Research has shown a link between STEM education on one hand, and literacy skills and language learning on the other. One of the interviewees on my channel is a Haitian American who spoke no English when he moved to the United States at age 6; however, it was an understanding of, and interest in, STEM that allowed them to engage with other pupils through ideas which helped them develop their language skills.

3. *STEM is only for certain types of people.*

For a long time, STEM has been dominated by males, and in the West, it has been white males. It used to be seen that STEM was not for girls, and equal opportunities were not available for Black and Minority Ethnic (BME) groups. This tide has been turning, with more prominent females and ethnic diversities in public STEM engagement, and campaigns to promote STEM as being for all people; *because it is*!

4. *STEM education only happens in the classroom.*

A lot of STEM learning can and does take place at home. Technology has made it possible for me to produce videos showing STEM demonstrations that people can easily replicate at home and do on a regular basis, not just within my country but globally. Various mobile apps can also be accessed for STEM learning at home.

5. *STEM is only for older students.*

In fact, early STEM education tends to lead to ongoing academic success and has pupils better equipped to cope with technology-focused require-ments when entering the workplace. Children are naturally inquisitive about the world, and sometimes this can be squashed by the impatience of adults fielding a million and one questions from their children. I know what this can be like; I have such a toddler at home. However, when I see him build-ing towers with his blocks, or when he asks me where his food goes, I am seeing his interest in STEM through his inquisitiveness, and I give him the tools to try different ideas for more secure towers, or I tell him what happens to his food once he's eaten it (leading to some very amusing comments). This inquisitiveness should be encouraged at all ages.

7.5   "HELLO AND WELCOME TO *STEM WITH MR N*"

In the 2019–20 academic year, my remit within my primary school was to deliver STEM education to all classes from Primary 2 to Primary 7 (6- to 11-year-olds) and develop the resources available within the school. In early 2020, I welcomed my first child into the world; then, 10 days later, the UK welcomed their first lockdown due to Covid-19. I was still responsible for the STEM education for the school, so I had to come up with a plan for how I was going to continue their STEM education.

Throughout the year, I had been looking on YouTube at times for ideas and inspiration for experiments. I quickly discovered a lot of STEM videos which showed interesting experiments but did not actually explain the science behind what was happening. To me, it is important that children get not only the "WOW!" moment of performing experiments, but also the "HOW" element.

I decided the best way forward was to create a YouTube channel and show the pupils some experiments they could easily do at home, while explaining the science behind what was happening. I shared the link with some other head teachers I knew to share with their schools. I honestly thought I'd get five views and never make another video. I was wrong.

That is when I became *STEM with Mr N*, first on YouTube and then on other social platforms. That first video proved very popular with pupils, so the videos continued during lockdown. This started as showing weekly science experiments and explaining the science behind what was happening. Very soon after starting, I was asked to partner with a couple of companies on giveaways, shortly followed by a robotics company who offered me a robot if I could use it in a couple of videos.

Over three years later I am still going strong, mostly because I see the impact STEM has on pupils. I have continued with science demonstration and explanation videos; however, my range of videos has expanded. I have enjoyed multiple partnerships with robotics companies, leading to an ongoing Robot Review series; I have a series of STEM Career Interviews where I speak with a diverse range of people working in STEM (more on this later); a series of videos exploring 10 Things You Should Know About different topics; and there are more ideas on the horizon at the time of this writing.

Creating these videos has given me even greater confidence when delivering STEM education directly to pupils in school, and the pupils in turn are excited to have a teacher who is active on their favourite social media platforms. This increased professional recognition, enthusiasm and confidence has led to me presenting termly science shows in my school and a lot of activities in my classroom (leading to some viral social media moments). It has also helped in developing a progressive STEM program for teachers in my school, and I am a recipient of the UK-wide Primary Science Teacher Award (2021–22) from the Primary Science Teaching Trust (PSTT), which recognises the exceptional efforts of teachers who make a difference to children's learning and enjoyment of science.

7.6   GETTING CHILDREN EXCITED ABOUT STEM

As I mentioned earlier, children and young people are naturally inquisitive about the world around them, and STEM is a way of finding answers to their questions. Digital technology has made it easier than ever to find information about any topic you desire, but STEM should be about exploration and discovery, not just looking something up online.

As a teacher, I have a curriculum I need to follow and teach, but I also want my pupils to be excited about what we are learning and the ones who actually want to push the learning forward. When I've been working with children, I've often found that, before getting fully involved in a topic, there needs to be a hook, something which makes the pupils excited and wanting to explore different topics. Advances in technology allow me to be able to explore more topics with my pupils in more detailed ways, but they still need that excitement. As Neil deGrasse Tyson said, "There is no greater education than one that is self-driven."

I often look to start my school year with a story which has a STEM focus. Over the past few years, I have used: books with a space theme; one where the multiverse was the main plot point; a book of poems about a character's STEM explorations; and a book about lost robots gaining consciousness. Stories with a STEM theme naturally pique the interest of children and lead to questions, which allows the teacher and pupils to move forward in discovery together, and there has been some early research showing that stories help young girls become engaged in STEM.

In the section about STEM myths, I highlighted the importance of STEM education starting when pupils are young, however in my experience the majority of primary teachers do not have a STEM background and therefore are not confident tackling this area of the curriculum.

This is another area where I use *STEM with Mr N*; to help support teachers with the delivery of their own STEM education in their establishments. All my science experiment videos explain the materials needed, possible alternatives, the process and, as I've mentioned, the science about what is happening in child-friendly language. I also tend to mention different ways the experiment could be expanded, encourage creativity and, where appropriate, mention careers that utilise the knowledge being gained in the video. All of this is about raising the quality of the delivery of STEM education so that our children and young people – boys and girls – are better equipped for entering an ever-increasing STEM workforce.

However, there is another issue at play, and that is childhood awareness of STEM careers, particularly for young girls, which is another area I try to target.

7.7   STEM CAREER INTERVIEWS

There seems to be some debate as to whether careers should be discussed in primary schools, or whether this is too early for pupils to be exposed to careers, instead following the mantra of "let kids be kids" and avoiding the topic.

When I was being assessed for the Primary Science Teacher Award 2022, my classroom was visited by two members of the Primary Science Teaching Trust (PSTT). My class was excited to be visited by *real* scientists. Following their visit, one of my Primary 3 (Year 2) pupils said the below quote to their mum. At face value, this could seem like an innocent statement made by a 7-year-old; however, it is a symptom of a wider issue. "There were two scientists in my class today and none of them were men!"

A report launched by Education and Employers in 2018, surveying 13,000 UK primary pupils aged 7 to 11 about their career aspirations, found that children start to rule out career options at an early age, and that there are multiple factors which affect the decisions about their future careers, including gender stereotypes, socio-economic backgrounds, lack of role models, and representation in media.

I am a good educator, and my work in schools along with my videos have been helping to raise science capital, which was noted in my own school by my head teacher at how much the pupils have developed through the work I have been doing.

However, with career aspirations starting to form at such an early age, I felt that as well as my usual videos, children should have the opportunity to learn about different career possibilities, and encounter people who work in these roles, regardless of their gender, background or location, while still in primary school to help broaden their horizons as they progress through their educational journey. I also think it's critical that teachers and parents be made aware of the changing career landscape and options; otherwise many children will be limited, especially those from marginalized groups. I felt I could also start to make a difference in this sphere, too.

One way to easily introduce career discussions in the classroom is to link the knowledge and skills being developed with those who require these for their careers. For example, here are some ways to associate classroom topics with careers:

- Numeracy and Mathematics – I teach about contexts in which budgets would be used like a financial accountant.
- Literacy – I teach about functional writing which can be used for a career in journalism.
- Physics – Specifically, when teaching about refraction during STEM time, I talk about lens technicians.
- Computing – When teaching about algorithms, sequencing and debugging in computing class, I introduce the world of software engineering and coding.

Role models also play an important role in raising the career aspirations of children. One of the key things I do as a STEM communicator is produce monthly interviews with people working in different STEM careers. This exposes children and young people to a diverse range of careers they may not have known existed, along with the diversity of ethnicities, backgrounds, genders and routes into STEM that exist. By hearing from a woman talking about her career in engineering, a Haitian who couldn't speak English when he started school in America, an IT director who quit university to take up an apprenticeship and a STEM program leader who quit an apprenticeship to go to university, our young people are exposed to people they may be able to identify with, to careers that speak to their interests and to opportunities they did not know existed.

When I decided to start the interviews, I honestly did not know where to begin. I decided to post on LinkedIn, Facebook and Twitter to see if anyone was interested. However, I also took the bold approach of cold-emailing people in STEM who I admired or had connected with on LinkedIn and found interest in what they were doing. I was delighted with some of the notable people I was able to interview, even being privileged enough to interview the world-leading forensic anthropologist, Dame Prof. Sue Black. The reason for all of the positive responses, and for people being willing to participate, is that they share the same view as me: that it is important to promote the diversity within STEM to young people.

There should never be barriers to people looking to enter into any STEM field, and as an educator and STEM communicator, I have the power to remove those barriers by developing different skills, introducing career discussions in the primary school to tackle gender stereotypes, expose children to different careers and introduce role models. Ultimately, however, the choice is there for children to make while we help them discover the passion that lights a fire inside of them.

7.8   WHAT CAN YOU DO?

When we tend to think of people who have made a change in the world, we think of great names etched into the collective consciousness, like Abraham Lincoln and Charles Darwin (born on the same day, 12 February 1809), or Stephen Hawking and Albert Einstein. However, most of the change that takes place in the world is from 'nameless' people who are changing their society from the inside.

I am very privileged that I hold a position in society that provides me with the opportunity to shape and inspire young minds, and also that I have the knowledge, skills and opportunity to make videos that promote STEM education, careers and fascinating information.

But you can also make a difference in the world. Talk to your children and the young people around you about the amazing wonders of STEM, explore answers to questions, try experiments, talk about careers and be creative. There are lots of opportunities to promote STEM in our everyday lives, and there are a lot of resources out there, like my own, which can help you.

'Teacher' may be my job title, but we can all be teachers by engaging and inspiring those around us.

ABOUT THE AUTHOR

Stuart Naismith is a primary teacher in Scotland, as well as a STEM communicator across social media platforms. He has a passion for engaging people of all ages in STEM education, but especially in explaining big topics in a child-friendly way, accompanied with practical activities to allow people to personally explore the concepts they are learning about. Stuart has undertaken training courses with the UK Centre for Astrobiology, the European Space Agency, the Royal Observatory in Edinburgh, and the Scottish Schools Education Resource Centre (SSERC), and is a Fellow of the Primary Science Teaching Trust. Stuart is married to a successful doctor, Viv, and together they have three beautiful children: a three-year-old son, James, and identical twin daughters, Sophie and Olivia, who are nine months old. When he's not teaching, Stuart is spending most of his time with his family, enjoying all the experiences that come with parenthood.

In the very little spare time he has left these days, Stuart loves reading, both fiction and non-fiction, and he has an impressive, and ever-increasing, collection of signed books by some of the best science communicators, as well as a number of fiction authors. Stuart also plays ice hockey, which he has been doing since he was five years old; a passion he shares with his father, who he plays on a team with. You can find *STEM with Mr N* at youtube.com/c/stemwithmrn and by searching for "STEM with Mr N" across social media.

8 Increasing UPTIME
The Journey from Banking to Engineering

Jenny Ambler
UPTIME Consultant Ltd, United Kingdom

8.1 EMBODYING THE HUMAN CHEQUE PROCESSORS

Leaving school at 18 with no idea of what I actually wanted to do in life found me working in one of the big four high street banks. Back in the 1990s, people actually still walked into banks every day to carry out their business. Everything was still very much paper based, so I found myself part of a team in the back office of a large city bank in the 'machine room'.

Thousands of cheques required processing every single day. Racks of these paper slips were delivered to our work stations, where each amount was keyed into a manual terminal and the cheque processed.

Accuracy and speed were required to get the massive amounts of paper transacted and, most importantly, balanced by the end of the day. To fail to do so meant a whole world of pain as the input at fault had to be found.

The early data programmers back in the 1950s were mainly women and at NASA were known as 'Computers'. You could say at the bank we were the human equivalent of the cheque processing machines which came later on that used the first visual recognition systems.

It was a tiring, monotonous, boring job that became even more uncomfortable in the summer months as the machine room heated up with no air conditioning. There was also the responsibility of handling cheques for large amounts which, if they were input incorrectly, could end up with a customer being debited or credited the wrong amount. That could lead to serious trouble if you were identified as the culprit! At the end of the day, it just didn't seem worth it.

After five years, I realised I couldn't carry on doing this for much longer. One day, I had enough and decided to leave and travel abroad to see if I could work out what I really wanted to do.

8.2 SWITCHING FROM SKIRTS TO TROUSERS

Cutting a long story short, after four months of living in Turkey, I decided that returning home might reveal the answer to my question. Returning with nothing other than what I carried was quite a cathartic moment, it was as if I could now restart afresh.

DOI: 10.1201/9781032679518-10

I had an aunty who ran a pub down in London, and she offered me the opportunity to go learn the ropes and live there whilst I sorted myself out. I didn't think I could handle the full-on lifestyle of working in a pub handling customers all the time, so I started looking for a job. A friend from my banking days told me that there was a new factory that had just opened on the outskirts of the city. They were looking for people through an agency to work on packaging lines.

I didn't know anything about working in a factory, but I thought it can't be as bad as the machine room at the bank, and I needed to earn some money. It was an eye-opening experience. At the bank there was even a dress code, including skirts for women; wearing trousers was frowned upon and discouraged. How far we have come; at least the factory provided overall trousers and no skirts, a winning situation for me.

This was another repetitive tedious job, but in another way, the shifts were long and counting to 36 packets per box would even stop me going off to sleep once I got home! I had also elected to work regular nightshifts, which was again a shock to the body and in hindsight certainly not recommended.

After 18 months of packing thousands of boxes of Doritos and other snack foods, an internal recruitment advert caught my eye. The company was looking for its first new level of employees.

8.3 BECOMING THE FIRST FEMALE TECHNICAL OPERATOR

Up to then, there had either been Technicians, Facilitators or Logistics team members. Now, they were recruiting for Technical Operator roles that would work closer with the technician group in operating and maintaining the packaging machines. With the encouragement of a few of the Technicians on shift, I was persuaded to put my application in and move to the other side of the packaging operation, running and maintaining machines.

Assessment centres were held and I made it through to the interview rounds where we were put through the PepsiCo Targeted Selection process that they used to recruit the right fit for the role. This was looking for people who were adaptable and able to pick up new skills whilst showing some resilience. These were held off site at a local hotel where we were put through interviews, group sessions, role plays and a technical practical test that required spatial recognition and dexterity.

A few weeks later, I learnt that I had been successful, becoming one of the first Technical Operators on site and the only female member of that select group! Now it was off for two months of onboarding and training, with two weeks spent at another production plant to gain further insights and some experience.

I had found something I liked doing! This was working with my hands along with thinking about overcoming problems. At the same time, there were still some downsides, and one of those was being the lone female in the group.

Over the next couple of years, I operated and maintained packaging machines on one of the two Doritos lines. Then a new product line was commissioned on site, and I joined that team to bring that brand to the market. Later on, that production line was extended with different models and makers' machines. This meant more training to safely operate and repair them. Small running repairs with maintenance was now part of our skill set as the Technician group, which spread over more areas as the site grew with more production lines being added.

8.4 INCREASING UPTIME IN THE INDUSTRY

By 2015, I had enough of working 12-hour shifts, including days and nights. The body can take only so much punishment. Fortunately, a redundancy came around, and my partner Andy and I were able to leave the Walkers Snacks business together. I had served 18 years with PepsiCo, additionally with 18 months working for the agency, and Andy had completed over 21 years as an engineer, latterly running the Condition Monitoring and Lubrication program for the site.

We were free to do whatever we wanted, so a plan was hatched to form an engineering consultancy based upon our experiences and skills built up over those years. Andy had already been busy planning the launch of UPTIME Consultant Ltd and a couple of months after we left in June 2015, we were in business.

Now, all we needed were paying clients. But first, we took off on a year of extended breaks and holidays with part of our severance pay, all whilst planning what UPTIME Consultant would look like.

We are a good team, as Andy would do the 'doing' whilst I covered as co-director and company secretary with the financial side as well as proofreading his thoughts, articles, offers and strategy. The first couple of years were challenging, as we had to build confidence with potential clients to get the ball rolling.

Now, eight years into the journey, we have a thriving business. We survived the financial bump of the lockdowns in 2020–21 partly by working for our main client in the food industry. You guessed it – PepsiCo is our number one client, although we have worked across industry sectors including automotive, water, gas network and many other SMEs in diverse sectors.

Who would have thought that the 18-year-old who didn't know what to do would end up as the co-director of a successful engineering consultancy? I'm certainly pleased I took the leap out of banking, as I don't think I would've still been in that industry now.

I'm still not sure of what I want to do, but then I can't understand why you have to pigeonhole yourself to one particular thing. I believe people are different and will find their own way.

Many thanks for reading a little of my personal journey. I hope it inspires you to consider making a change, taking a leap of faith, or maybe even a career in engineering or technology. ,

ABOUT THE AUTHOR

Jenny Ambler is the co-director of engineering consultancy UPTIME Consultant Ltd based in the UK. A true inspiration to other women to join the industry, she shares her birthday on 8 March with International Women's Day. She supports the Breast Cancer Charity 'Walk the Walk' by raising money and taking part in their annual

walking challenges over marathon distances – imagine up to 15,000 women walking 26.2 miles starting at midnight. Over the past few years she has completed the
London, Edinburgh and Icelandic editions of 'Walk the Walk Moonwalk' and this
year will be working as an event volunteer in London to give her feet a well-earned
rest! Jenny still enjoys working with her hands by crafting with Paracord and making
Amigurumi figures.

9 Unexpected Paths, Endless Possibilities

A Journey of Integration between Literature and STEM

Corey Marie Hall
STEM Education Works, USA

Science and literature are not two things, but two sides of one thing.

—**Thomas Huxley**

As the Director of Curriculum Solutions at STEM Education Works, my role revolves around bringing fun and exciting STEM experiences to K-12 students within the age range of 5–18 years. I have the privilege of working with cutting-edge technologies such as robotic arms, microcontrollers, augmented reality, and 3D printing, all with the aim of preparing today's students for the careers of tomorrow in industries like manufacturing. However, this path I find myself on was unexpected, and STEM was not the focus of my studies or even my passion for most of my life. In this chapter, I want to take you through the unexpected journey I embarked on, from being a bookworm to becoming a leader in STEM. Along the way, I hope to shed light on how those of us from creative and literary backgrounds bring a richness and depth to the world of STEM that has often gone unnoticed.

9.1 PAGES TO PROTOTYPES: A BOOKWORM'S EXPEDITION INTO THE WORLD OF STEM

I've always been the bookworm, with my nose buried in captivating stories. I would lose myself in the pages, reading late into the night, even hiding under the covers with a flashlight. Throughout school, I was the kid who would eat lunch with a book, spend mornings in the hallway engrossed in a novel, and get genuinely excited when I knew it was library day. I was considered a good student and even part of the gifted program. There was no doubt that I was smart, but it seemed like there was a specific kind of intelligence associated with that label. This belief shaped my decisions from an early age, directing me towards pursuits aligned with my perceived strengths.

DOI: 10.1201/9781032679518-11

During my middle and high school years, I excelled in English and writing. I gravitated towards courses that focused on reading and research, such as European history, American literature, and business writing. While I did take the required math and science courses, as I progressed in my education, I began to adopt a learned helplessness mentality towards these subjects. I dropped out of honors classes in math and science, convincing myself that I lacked the capability to successfully tackle advanced coursework in those fields.

It wasn't until I entered college in the 1990s, a time when the Internet and computers were just beginning to make their mark, that my understanding of my strengths and interests started to shift. In my very first semester, I enrolled in a course called Introduction to Computers, held on Monday nights from 7 to 10 p.m. It was an unpopular time slot, mostly filled with freshmen who couldn't get into their preferred classes and a group of elderly residents auditing the course as part of a community outreach program.

At first, I approached the class with little interest and a negative attitude. But as I delved into the world of floppy disks and WordPerfect, something unexpected happened—I developed a genuine interest in technology. While I didn't care much for the hardware or programming aspects, there was something about the way computers could efficiently organize and manage information that captured my attention.

This newfound fascination led me to explore different fields and disciplines during my college years, resulting in multiple changes to my major. I initially pursued a degree in nursing, then shifted to business, and even considered criminal justice at one point. However, it was during my sophomore year that I discovered my true calling: teaching.

Recognizing the power of education to inspire and shape young minds, I eventually settled on a double major in middle school science and social studies. This decision allowed me to combine my love for literature and the humanities with my growing passion for technology and STEM education, creating a unique and fulfilling path that continues to resonate with me to this day.

9.2   CULTIVATING CURIOSITY: HOW A MICROBIOLOGY CLASS SHAPED MY APPROACH TO EDUCATION

I experienced a pivotal moment in college that solidified my passion for inquiry-based learning experiences and shaped the kind of teacher I aspired to become. It occurred during a microbiology class, a subject that had always been challenging for me. However, I was fortunate to have a professor who had a unique approach to teaching and assessment. This professor, unlike any I had encountered before, did not believe in traditional grades. Instead, he valued the process of learning and mastery of concepts. He created an environment through which we were given endless opportunities to experiment, explore, and refine our understanding of microbiology. It was a shift from memorizing discrete facts to embracing a deeper level of comprehension through hands-on experiences.

In this class, I felt truly empowered as a learner. I was encouraged to ask questions, think critically, and take ownership of my education. The professor's emphasis on inquiry-based learning sparked a genuine curiosity within me, motivating me to explore beyond the confines of the curriculum. I realized that true understanding came not from regurgitating information, but from engaging in meaningful exploration and discovery.

This experience in the microbiology class profoundly influenced my educational philosophy. It taught me the importance of fostering a love for learning and providing students with the opportunity to engage in meaningful, hands-on experiences. I wanted to create an environment where students could explore, experiment, and learn from their failures as they strive for mastery.

This professor showed me that the true essence of education lies in empowering students to become lifelong learners, fueled by curiosity and a genuine passion for knowledge. He instilled in me a belief in the transformative power of inquiry-based learning and the importance of creating meaningful learning experiences for students. I carry this experience with me as a constant reminder of the impact that passionate and dedicated educators can have on shaping the lives of their students.

9.3 FROM NOVELS TO CODE: EXPLORING THE INTERSECTION OF CREATIVITY AND TECHNOLOGY

These newfound interests in technology and science sparked a curiosity within me. As a newly minted middle school teacher, I began to explore different software applications and digital tools that could enhance my productivity and streamline my work. As I dived deeper into the world of computers, I discovered the power of coding and its ability to create solutions and bring ideas to life.

Driven by my growing fascination, I decided to take a leap of faith and delve into the world of coding. I enrolled in programming courses and immersed myself in learning various coding languages. HTML and JavaScript quickly became my allies, as I grasped the fundamentals and started creating my own classroom web pages and interactive elements. The sense of accomplishment and satisfaction I felt when my code produced desired results was invigorating.

Through this transformative experience, my journey from being a bookworm to a leader in STEM gained a new dimension. It became not only about my personal growth, but also about the impact I could have on the lives of my future students. I wanted to inspire them to approach STEM with the same sense of curiosity and inquiry that I had discovered.

As I continued to explore technology and coding, I realized that these fields were not just tools but gateways to limitless possibilities. They provided opportunities to blend creativity, critical thinking, and problem solving, all of which were essential for fostering a love for learning and creating an inclusive and engaging STEM education. I realized that the connection between my love for literacy and the world of STEM was stronger than I had ever imagined. Technology and coding were not merely abstract concepts, but tools that could enhance and elevate the power of storytelling and information sharing. I recognized that by merging my passion for literature and my newfound skills in technology, I could open doors to new possibilities and inspire others to embrace the relationship between creativity and STEM.

9.4 FROM TEACHER TO MAKER LIBRARIAN: EMBRACING STEM IN THE LIBRARY

Motivated by this realization, I made a pivotal decision to transition from my role as a classroom teacher to a STEM-focused school librarian, spearheading the establishment of a vibrant makerspace brimming with cutting-edge technology and resources. Equipped with 3D printers, robotics kits, virtual reality tools, and a plethora of building supplies, our makerspace became a hub of innovation and exploration. This leap into uncharted territory was fueled by my unwavering determination to merge my passion for literacy with my burgeoning expertise in STEM, with the ultimate goal of providing students with a holistic and engaging learning experience.

As a maker librarian, I eagerly embraced the challenge of incorporating STEM principles into my classroom and beyond. Collaborating with teachers across various content areas, I worked hand in hand to develop cross-curricular projects that integrated STEM concepts into diverse subjects such as language arts, social studies, and even physical education. By fostering these interdisciplinary connections, students were able to see the practical applications of STEM in different aspects of their education, reinforcing the importance and relevance of these principles in their lives.

In addition to hands-on experiments and activities, I recognized the growing importance of coding skills in our increasingly digital world. With enthusiasm and dedication, I taught students the fundamentals of coding, guiding them through interactive coding exercises and encouraging them to create their own programs and digital solutions. Through coding, students not only developed computational thinking and problem-solving abilities but also gained a sense of empowerment and the confidence to tackle complex challenges.

Beyond the traditional classroom setting, I also facilitated a student-led Dungeons and Dragons club within the library. This unique extracurricular activity provided a creative outlet for students to develop their storytelling, teamwork, and critical thinking skills. By immersing themselves in the world of Dungeons and Dragons, students were able to apply their STEM knowledge in imaginative and collaborative ways. They learned to design intricate game worlds, devise complex puzzles, and use logic and reasoning to navigate through their adventures. Witnessing their growth and engagement in this student-led club was an extraordinary experience, showcasing how the principles of STEM can be seamlessly integrated into recreational activities to cultivate a love for learning and exploration.

9.5 FROM MAKER LIBRARIAN TO STEM CURRICULUM INNOVATOR: SHAPING THE EDUCATIONAL LANDSCAPE

After dedicating several years to my role as a maker librarian, fate took an unexpected turn with the onset of the COVID-19 pandemic. Like many others, I found myself facing the harsh reality of layoffs and the disheartening impact it had on the educational community. Despite the challenges, this setback became an opportunity for me to reflect on my purpose and aspirations as an educator.

Driven by a deep desire to expand my reach and impact more students, I embarked on a new chapter in my career. The experience I gained as a maker librarian, collaborating with teachers and witnessing the transformative power of STEM education, fueled my determination to make a broader difference in the field. This led me to my current position as the Director of Curriculum Solutions at STEM Education Works.

In my role, I am privileged to work alongside a passionate team of educators and experts who share a common vision: to bring innovative STEM experiences to students across the United States. Together, we strive to develop curriculum materials that not only meet academic standards but also inspire and empower students to become lifelong learners and confident problem solvers. Our aim is to create meaningful and engaging learning experiences that ignite curiosity, foster creativity, and nurture a love for STEM.

I have come to recognize the immense value of individuals with creative and literary backgrounds in the world of STEM. The ability to think critically, communicate effectively, and approach problems from unique perspectives is a vital asset that enriches the field and promotes a more inclusive and multidimensional approach to learning. By embracing the diverse talents and backgrounds of educators, we can create learning environments that cater to the strengths and interests of every student, allowing them to thrive and excel in STEM fields.

With each passing day, I am grateful for the opportunity to make a difference in the lives of students on a broader scale. I am committed to pushing the boundaries of STEM education, fostering creativity and innovation, and empowering the next generation of thinkers, makers, and problem solvers. Together, we can shape the educational landscape and inspire students to embark on their own remarkable journeys in STEM.

Through this chapter, I hope to inspire others who, like me, may have once believed that their passion for literature and creativity had no place in the realm of STEM. I want to emphasize that it is precisely these diverse perspectives and skills that have the potential to revolutionize the way we approach STEM education. By embracing our creative instincts and leveraging technology as a tool, we can unleash the full power of STEM, transforming it into a dynamic and inclusive field that not only fuels innovation but also nurtures the innate curiosity and imagination within all of us. Whether you are a student, an educator, or someone simply curious about the intersection of literature, creativity, and STEM, I encourage you to embrace the synergy between these fields. Let us celebrate the richness and depth that individuals from creative and literary backgrounds bring to the world of STEM, and together, let us shape a future where imagination and innovation coexist harmoniously.

ABOUT THE AUTHOR

Corey Marie Hall, PhD, MLIS, MS, is an experienced educator, librarian, and college professor. She is passionate about kids, literacy, and technology. In her position as Director of Curriculum Solutions at STEM Education Works, she has the opportunity to work with other amazing educators to create and edit STEM curriculum for K-12 students. In the process, she gets to experiment with new technologies and help educators implement them into their own teaching. She has a PhD and MS in educational technology and an MLIS with a specialization in emerging technologies.

10 Maneuvering the Moving Sidewalk of Your Career

Susan Lubell
Steppe Consulting Inc, Canada

10.1 DEFINING THE STAGES OF OUR CAREERS

The world all around us and how we live our daily lives are constantly changing and advancing, even more so when it comes to advances in technology. Our careers, in any field, are never a linear path from point A to point B; there are always many detours, branches and routes to explore along the way. These diversions naturally shape our careers, and how we react to the opportunities that present themselves will inevitably have a profound impact on our long-term direction and career success. The Robert Frost poem – "Two roads diverged in a yellow wood … and I – I took the one less traveled by, and that has made all the difference" – springs to mind.

To bring substance and breadth to the discussion on empowering women in STEM, I reached out to other women engineers from university students to those well into their careers with over twenty years of experience to see what questions they would ask if this were an interview on STEM careers, specifically careers in Engineering. It was interesting to see the broad overlap in questions – no matter where in their own career paths these women fall. Their questions form the basis for answers that my younger self might pose to me. If our career paths are wide open and unknown, we nevertheless need ways to consider and navigate the path forward – how to hop on the moving sidewalk of a career and stay balanced as it accelerates.

Moving into future generations of technical and innovative evolution, the roles that we will hold in our careers may not yet even exist; we need to be open to this possibility. There is no question that this evolution is happening, and the rate of change is increasing. The road into STEM careers is no longer a linear path from high school math and science, to university or college, to a progressive career path with a single employer. This is what makes the future so exciting – there are wide webs of career paths in which to apply an interest and aptitude for STEM. The opportunity to step on and off the moving sidewalk occurs many times throughout our lives.

Our careers hold three distinct phases. The early career phase can be characterized as learning and exploring. The mid-career phase is more focused on demonstrating competence, implementation of know-how, and all-round performance. The final career phase can best be described as giving back – to your employer, your broader community and to individuals or groups as they pursue their own careers.

DOI: 10.1201/9781032679518-12

10.2 EARLY CAREER – LEARNING AND EXPLORING

Like many young women in high school, I was 'good at' and interested in math and science. High school career day programs had at least one parent, usually a dad, who came to talk about careers in engineering and invariably, they talked about careers with the local electrical utility or car manufacturer – a local industry. Neither of these end goals had any appeal to me as I was exploring various career options, so I ignored Engineering as a viable step forward. With a family background in medicine, my career sights were set on the healthcare field – most likely becoming a doctor or possibly a geneticist. I applied for and was accepted into the first year for a Bachelor of Science degree at university, got a summer job in a medical research laboratory, and started down that more common and narrowly defined STEM career path.

As I settled into my university education and worked summer intern roles in the field, it quickly became obvious to me that I preferred applied science over the pure science of working in a laboratory. Talking to classmates who were enrolled in undergraduate programs in engineering and a few professors, the career path I could see myself following was more aligned to chemical engineering, with its focus on problem solving and collaborative process improvement. This clearly was not my previous introduction to engineering; it wasn't all about electricity and cars! I was excited by the prospects.

The question of how to break into an engineering career was asked by more than one of the women Engineers who provided questions. The importance of varied summer work terms, co-operative education work terms, and internships in different industries, companies, and even different parts of the country is not easily measured. Particularly in a field as broad and diverse as Engineering, this work term experience is the first significant step towards identifying what industry and work environment is the right fit for you – an energy production plant, manufacturing, food processing, a research and development laboratory, a water treatment plant, or a head office in a large city all have varied expectations and cultures.

Certainly, it does not mean that you will permanently work at the organization where you complete your work terms, but these work experiences are the beginning of exploring your options. A benefit of pursuing a STEM career specifically in engineering is that internships and careers tend to be better paid than many other fields. This can be a substantial boost when paying for your education and looking at long-term financial stability.

Unfortunately, the prevailing thought of many employers remains that every Engineering graduate wants to focus on engineering design and operations work, with a heavy emphasis on performing technical calculations. This simply isn't true.

The field of engineering is broad and quite varied. I spent the first few years trying out different career options to explore what type of work would keep me engaged and interested. My summer work term roles in medical R&D laboratories, an electric utility testing facility and finally in a process engineering role at an oilsands production site, followed by a few years as an engineer-in-training working in oil and gas production, led me to a career path in industry. This early career phase is also the period when you may decide additional education or certification knowledge is required to advance your existing career path or alternatively, to head in a different direction – hopping on and off the moving sidewalk.

It quickly became apparent that more formal business knowledge was a necessity if I wanted to branch out from a traditional Engineering daily plant production focus, so I took the leap into completing my Master of Business Administration (MBA). My own career was not a straight path from a Bachelor of Science degree to chemical engineering to asset management. I am very grateful for the opportunity to work in oil and gas production in the Canadian north early in my career, in roles as a chemical engineer, project manager, reliability engineer, and in asset management. I gained a solid appreciation for the challenges of the work environment, and the way that various teams and engineering specialities can work together to solve problems. Now when I work with clients or teach about strategy, planning and implementing asset management, I can truthfully say that I've been there and done it!

This early phase of your career is also the time to move around the country, try out various operating roles, and work those job roles with odd hours and shift work. This experience builds the foundation for your career and is invaluable when, later, you are making decisions that impact others or considering technical options. A few middle of the night shifts outside in a chilly −40°C brings a new perspective to how long it can realistically take to complete activities in the winter or the impact of weather on asset operations, equipment performance, and production.

10.3   MID-CAREER – COMPETENCE AND PERFORMANCE

By the mid-career phase, it is typical to have specialized into a specific industry – gaining in-depth knowledge of the technical and regulatory context and to be taking on additional responsibilities in your career. This is the ideal time to be on the lookout for career opportunities as they arise. Consider those front-line leadership roles, identify what specialty you want to focus on, and be open to moving between teams or even organizations to find your niche. As noted earlier, the evolution of the STEM field and the rapid and constant changes in technology make continued learning a necessity to build and sustain competency.

The idea that your direct supervisor is the sole source of advice for career advancement and to discuss career options is outdated. The pace of technological change and innovation is advancing quickly, and many of the future roles for engineers haven't yet been defined. Part of the challenge of your mid-career is to narrow in on the aspects of the engineering work that inspire you to give your best, while not limiting yourself. If an as-yet unknown fantastic career opportunity pops up, it's important to have the up-to-date skills and expertise to pursue it.

10.3.1   WHY DO MANY ENGINEERS LEAVE THE FIELD MID-CAREER?

Many engineers, both women and men, choose to leave the field at this stage of their career. The problem-solving and collaborative approach gained through the early phases of an engineering or STEM career will serve you well in the next stages of your career, whether you shift into a related field within your industry or take a leap in a completely different direction such as following an entrepreneurial path in another industry.

Several questions from the women engineers hinged on the challenges of balancing your own career with a partner, likely two careers, and possibly children. It's no great surprise that many women engineers marry other engineers, as the research conducted at the University of Calgary confirms – this is who you are likely to meet in school, through mutual friends, or even at work. If your partner is also an engineer, decisions need to be made that will affect two professional careers. This is always a tricky question, and it never fails to arise in every discussion on career advancement when women engineers get together. It's even trickier if some of the options include moving cities to advance a career path.

Each person and each family will need to navigate the options in the way that works best for their own lives. My own family balanced my engineering career advancement, my engineer spouse completing a PhD with an associated career path, and welcoming a child into our family. There is no one size fits all – personal experience is that we tried various options over many years to balance two growing professional careers with the strong desire for a stable family home life. The best option wasn't a single path; there were detours, side tracks, and various points where one partner's career advanced faster than the other along the way. Communication and common family goals go a long way when the road gets bumpy at times and difficult career-balancing choices need to be made.

10.3.2 The Importance of Career Guidance and Mentorship

Another common theme of the questions was about the people who provide career guidance, mentorship and simply those who believe that you could take on a more challenging role or were ready for additional responsibility. These may be peers and supervisors inside your own organization, colleagues, and friends in similar stages of their own careers outside of your organization, and others at the third stages of their own careers that you may have met through previous career roles, industry associations, or similar career networking opportunities.

Throughout my career, I've had supervisors and colleagues who believed in my abilities to contribute, who encouraged me, and who opened the door to other options by asking me to work on new projects or take on additional responsibility. I have also had supervisors who tried to restrict my career growth and limit my contributions. I have learned from both types of supervisors and consequently developed my own style of how I treat others, and the encouragement and opportunities that I provide to direct reports and others in the profession whether they are on my direct team or not.

10.3.3 The Increase in New Opportunities

In the mid-career phase, new opportunities for increased technical specialization, management, and leadership are likely to arise. By this point, you should have a good idea where your natural inclinations and aptitude lie. A key part of this phase is being ready to leap on opportunities when and where they arise, even if they don't look like a direct path. Sometimes leading a challenging improvement project or actively participating in a large software roll-out initiative can be a basis for demonstrating your ability to manage people, make strategic business initiative progress within budget constraints, and engage the organization in a major change initiative, not to mention the ability to gain valuable technical skills and stay current with the latest technology developments.

When spotting a related opportunity, the best action is to jump on it and explore it rather than hanging back and waiting for a direct career path linear progression that may never come. At the midpoint in my career, I was delivering a lot of internal technical training within my company for staff located across several Canadian provinces as part of specific asset management, reliability, and maintenance improvement initiatives. When the local college asked my direct supervisor and other industry specialists who they would recommend delivering training as part of a national certification program, my name was put forward, and I took on this additional career opportunity. I have now been delivering national certification training for over fifteen years and continue to network with other practitioners in the field this way while giving back to others as they advance their own careers.

10.3.4 THE CRITICALITY OF NETWORKING

The importance of networking cannot be underestimated. Many career advancement roles within organizations are filled based on demonstrated competency – knowledge, skills, work experience, aptitude, and attitude – plus an expressed interest in taking on new roles. It's also a matter of being top of mind when organizations are trying to fill roles. So many engineers, and their supervisors, move between roles and organizations that it's very possible to be in a single role where your supervisor has changed several times and they may not be aware of your career interests or aspirations. Opportunities may arise at other organizations and, should this be the direction that you want to pursue, it's extremely helpful to have a colleague at that organization that you can reach out to learn more about the culture.

I've navigated my career through five major downturns and countless economic blips in my industry and the economy at large – the ability to network and land on my feet at another organization has been critical for me. In the energy sector, the idea of joining a single organization for an entire career of forty or more years is simply not a reality. Mergers, acquisitions, and divestments of entire companies or divisions are relatively common. On more than one occasion, I've found myself working with the same colleagues in three or more organizations over a fifteen-year period – it's a reality, not a coincidence.

This mid-career phase is solidly based on demonstrating competence in your chosen field and delivering consistent performance that advances you as an individual, aligns to your organization's goals and objectives, and supports an ever-growing network within your industry.

10.4 FINAL CAREER PHASE – GIVING BACK

By the time that you reach the final stage of your STEM career, there is a good chance that you have been recognized as a specialist in the career path that you've chosen and are ready to give back. This may include formal or informal mentorship roles for earlier-career engineering professionals, the option to contribute to industry professional associations, or the ability to provide sage experience and technical advice in your area of specialization. It is exceedingly rare for anyone to work alone in the modern workplace, and the experienced professional likely finds themselves in an informal mentorship role even without specific managerial responsibilities.

As my own career within employer organizations has progressed, I have sought out the opportunity to advance outside of my organization through continued personal education and associated certifications, attending and speaking at technical conferences, delivering webinars and podcasts, and contributing my time to my industry not-for-profit professional association. Through this, I have had the opportunity to expand my network, stay current and up to date on the latest thinking in the field, contribute to content knowledge, and serve in elected national roles for the PEMAC Asset Management Association of Canada, including four years as President, and as the elected Chair of World Partners in Asset Management (WiPAM).

I frequently discuss career options with women and men who are at earlier stages of their own career paths in STEM, specifically engineering. Many want to discuss their choices and the pros and cons of the opportunities in front of them so that they can make more informed decisions. At this third stage of my career, my path has shifted, from progressive employee roles and taking on higher levels of managerial and leadership responsibility, to one where I now provide consulting and advisory services particularly for asset management teams in transition as they seek to improve their own performance.

10.5   THE ELEPHANT IN THE ROOM

More than one of the questions from the women engineers centres on the role of women engineers and the unique gender bias aspects of this career choice. It would be remiss not to address this in a book focused on empowering women towards future careers in STEM. While the field of engineering is becoming more diverse and more women are entering the field every year, the number of women in engineering is still comparatively small. It is not uncommon for me to be the only woman in the room at a senior management meeting or to instruct an industry certification course and have only one or two other women in the room.

This can be a challenging position when you are distinctly noticeable in the room. It requires a level of confidence in your own abilities to be able to speak up and be heard in a work environment where many others are competing to be seen and heard, or where work relationships are furthered in the locker room while playing sports. This may not be the case in all fields of engineering or industries; however, in the resource energy sectors, this continues to be the norm. Learned skills such as public speaking, and the very real need to be more outgoing in a business situation even if this is not your personal comfort zone, support career advancement.

This is one area where I strongly believe that visibility of women who are already in the field of engineering, developing a network of women engineering peers both inside and external to your own organization, and the support of both women and men to encourage young women to stick with careers in Engineering is paramount. This is part of giving back at this stage of our own careers as women in STEM. Substantial progress has been made in this area over my career lifetime, but this is an ongoing area where we must continue to encourage and inspire younger women for their careers in STEM.

Overall, it is essential to create a supportive and inclusive environment that encourages younger women to pursue their interests in STEM and specifically in engineering. By providing opportunities and support, we can help close the gender gap in STEM and create a more diverse and innovative workforce. Women are making strides in the field of STEM careers and engineering moving towards the critical numbers that will demonstrate a gender-balanced workplace that recognises the value of all to contribute. To conclude, be confident in your skills and your ability to contribute to advancing yourself as an individual, contributing to the success of others, to your organization and the wider community at this third stage of your career.

10.6   FINAL THOUGHTS

The theme of this book, *Empowering Women in STEM – Working Together to Inspire the Future*, has provided an opportunity to invite other women in engineering to contemplate their own career paths and to reflect on my own career as I reach the third stage of giving back. As technology continues to advance at a rapid pace, affecting every aspect of our lives, it is an especially exciting time for women who are already on a STEM career path to consider their roles and opportunities to shape the world around us. With an overall perspective of three career phases, and the considerations unique to each phase, it will be easier to hop on the moving sidewalk of our careers and stay balanced as the sidewalk accelerates.

ABOUT THE AUTHOR

Susan Lubell, P. Eng., MBA, MMP, CAMA, is the Principal Consultant, Steppe Consulting Inc, and author of *Root Cause Analysis Made Simple – Driving Bottom Line Improvements by Preventing One Failure at a Time*. She specializes in asset management and reliability strategy, cost-effective lean maintenance programs, and operational excellence. Susan's career has focused on production operations in the oil and gas, mining and energy sectors, as well as teaching maintenance and asset management college courses at the national certification level. Along the way, she has achieved her Professional Engineering, Maintenance Management Professional (MMP) and Certified Asset Management Assessor (CAMA) certifications and has held leadership roles within PEMAC Asset Management Association of Canada and WPiAM World Partners in Asset Management, authored a book and contributed to another. Susan has a keen interest in building competency and capability in others and brings over 25 years of progressive experience to drive asset management, reliability, and maintenance business decisions, and to implement improvement opportunities in asset-intensive production companies.

11 Exploring Innovation, STEM, and Entrepreneurship
A Journey of Learning and Growth

Viktoria Ilger
Creators Expedition - an AVL Initiative, Austria

11.1 MY JOURNEY AS A BUSINESS GRADUATE INTO THE STEM WORLD

When I reflect on how my career path has evolved over the past 10 years, I have to consider who I am and what truly sets me apart. I have not followed a linear career path so far, and even to this day, I cannot make a final decision on which direction to pursue. However, I do know that my journey thus far aligns well with my personality, because what defines me is my need for new perspectives and experiences, which are as essential to me as breathing.

Since I was a child, I have loved being in the company of others, engaging in conversations, and believing in family and friendships as an integral part of my life. I am not fond of following routines and often wonder why we can't just do things differently for a change! As a little child, I would negotiate with my parents to let me sleep with my head at the foot end of the bed and if they wanted to go left, I would always insist on negotiating to go right, even to today!

Many years later, during a lecture by a neuroscientist, I learned that challenging one's own brain by breaking routines and seeking new experiences strengthens creativity and even helps prevent dementia. When we repeatedly follow the same routines, our brains work efficiently, but there is a risk of falling into an "autopilot" mode. This can lead to decreased attentiveness and reduced ability to find creative solutions. By consciously breaking routines and exploring new paths, we force our brains to become more active. They must adapt to new environments, process new information, and potentially solve new problems. This "brain-active" behavior can promote mental flexibility and creativity.

This characteristic has stayed with me, and once I understood that it is an essential part of who I am, I was able to navigate toward a career direction that challenges and fulfills me – the field of innovation!

DOI: 10.1201/9781032679518-13

11.2   MY DECISION TO STUDY BUSINESS ADMINISTRATION AND MY ENTRY INTO PROFESSIONAL LIFE

I have fond memories of my school days, as they were filled with a lot of fun, and it was a place where I could see my friends every day. I didn't have any problems in school, performing well in many subjects in high school, but I didn't have a clear strength or preference for a particular field. There might have been a few subjects that I preferred, but it depended largely on the teachers and their style of teaching. I could quickly get excited about things, and I loved preparing and delivering presentations in front of the class, but I struggled with teachers who wanted to impose too many guidelines.

My favorite was the Italian teacher who simply explained that we could do our homework or not. He would check the work we submitted – that was the service he offered – and those who wanted could avail of this service. For those who didn't need it and could pass the exam without it, they were welcome to forgo submitting their homework. That was exactly the right approach for me; it was as if he understood me, and I believe Italian was the only subject in which I regularly submitted my homework. Plus, I still speak Italian fluently and happily!

Back then, I felt like what they call an "all-rounder". Both my parents had non-technical backgrounds, with a focus on economics and law, so technology was never a big topic for me. When I thought about what I would like to do, I tried to imagine what my daily life would look like in a career, and I quickly realized that I didn't want to specialize in just one subject. I wanted to choose something that would allow me the freedom to decide and specialize later on.

Business administration seemed to offer that freedom, and besides, I had always been interested in economics. However, when I started studying, it quickly became apparent that my expectations were wrong. I had imagined that it would bring out my entrepreneurial creativity, but it turned out to be quite different. The first semesters were characterized by dry business mathematics, accounting, macroeconomics, and microeconomics. To my surprise, I found myself enjoying these subjects immensely, while subjects that I thought I would like, such as marketing, interested me less.

Since the university allowed for the prioritization and sequencing of courses, I progressed quickly and became very ambitious. People around me repeatedly mentioned that although business administration was a classic field, it was becoming increasingly difficult to find a good job due to the high number of business graduates. This drove me to be even more ambitious, and I wanted to stand out from the crowd.

I pursued international experiences, aimed to improve my language skills, and completed as many internships as possible. In the master's program, I chose specializations that were particularly challenging and were not commonly chosen by my peers. Therefore, I decided on auditing and corporate valuation, as well as digital business models and information systems.

Both specializations brought me a lot of joy, although looking back, I can't really understand why I chose auditing and corporate valuation. I would be a total misfit as a tax consultant or auditor. Nonetheless, at the time, I thought it was a good choice. In hindsight, the ambition I developed during that time was beneficial, but it also distanced me from myself a bit. If I had taken a moment to reflect and consider what truly defines me, I probably would not have chosen that specialization.

After completing my studies, I wanted to find a job quickly and enter the professional world. My parents even offered me the opportunity to take a short break, but I didn't want to at that time. Driven by the desire to establish myself, I accepted a job as an executive assistant in a bank.

I wanted to explore the world of banking and thought that being an executive assistant would be the ideal job to gain a good overview and understand how C-level executives work. I imagined it to be highly responsible, incredibly educational, and extremely exciting. I must admit that it was all of that, but reality quickly caught up with me.

As an assistant to the executives, I never left the office before they did. I was responsible not only for preparing board and supervisory board meetings and other areas, but also for ensuring their punctuality for appointments and handling matters related to their personal lives. Although I was interested in the financial world, I became increasingly unhappy with my daily work. I had maneuvered myself into something that didn't align with who I am as a person, so I decided to completely change direction after two years.

Would I do it the same way again? Honestly, I don't know. Looking back, I can only advise any student entering the professional world not only to consider the relevant skills when choosing a job, but also to listen to oneself and consider whether a job truly aligns with one's mindset and personal qualities. Skills can be acquired, but our personality defines us, and in my view, if a job doesn't fit with who we are, it affects our personal happiness.

During that time, I gained these and many other insights, and to this day, I maintain an interest in the financial world. However, I don't think it's my place to change something in the long run. Furthermore, I wanted to do something completely different, and so that opened the door to a world that I believed had nothing to do with me at the time—the world of science, technology, engineering, and mathematics (STEM).

11.3 MY ENTRY INTO THE STEM WORLD

After completing my studies, my goal was to build a successful career. Spoiled by academic success, I thought that with enough ambition, it couldn't be that difficult to establish a career, whatever the meaning of a "career" may be. During my time at university, I wanted to work in a large company that would offer me numerous opportunities. However, since I didn't find happiness in my job as an executive assistant in a major bank, I considered what the alternative could be.

I wanted to explore a new industry, engage in a different role, and so I decided to work in a small company. When I arrived for the job interview, it truly felt like stepping into another world, and to be honest, I wasn't sure what to make of it at first. In contrast to what I had experienced before, the office seemed a bit more disorganized, people were not only casually dressed but also more relaxed in their interactions, and out of the corner of my eye, I noticed a table football spot. And there I was, a little overdressed, surprised that the CEO personally and effortlessly greeted me at the entrance and asked me what I enjoyed doing in my free time.

Honestly, this question overwhelmed me because I was prepared for everything but personal questions. However, I think it was precisely because of that question that

I quickly felt at ease. Even though it was a different world and I wasn't sure if this job was the right decision, I accepted it, and from that moment on I immersed myself in the world of software development.

In a small company with around 30 employees, everyone had multiple responsibilities. I was primarily hired to collaborate with clients in developing digital business models, which would subsequently be implemented in software development. Therefore, I had the opportunity to accompany these implementations and delve into the world of agile project management. The company consisted of two-thirds software developers; they all warmly welcomed me, and we quickly became friends.

The only thing that bothered me at the beginning was the feeling that they explained things to me in great detail, sometimes at the level you would use to explain it to an elementary school child. For this reason, and because I needed to learn better what was possible in business model development, I actively started familiarizing myself with the world of software development. I even attended two courses on basic principles of software development. I did this not because I wanted to develop things myself, but because I believe that having a general understanding is necessary, and the deeper I immersed myself, the more I realized how exciting the field is, especially the opportunities and leverage it offers when you begin to grasp technology and its possibilities.

This remains my perspective to this day. I don't believe that I necessarily need to be able to implement technical solutions myself, but it is important to understand what new technological advancements are capable of. What does a new development mean for us as individuals and for entire industries?

Within the company, this attitude was quickly received positively, and I was given numerous opportunities to learn new things and contribute to various projects. I learned the fundamentals of user experience design, began to understand firsthand the value of agile project management compared to traditional methods, learned to work with challenging clients, and took on the leadership of my first European research projects.

Furthermore, the company was going through a difficult phase and wanted to transition from project-based work to scalable products. They had recently completed several research projects on digital assistance systems with smart glasses and wanted to create a product based on the results. I learned what it meant for a company to be in turmoil, and I am still impressed by the unity it created within the team. There was a sense of excitement, and everyone wanted to contribute at that time. We didn't just spend time together at work but also gathered in the evenings with pizza, drinks, and games.

Unfortunately, I also experienced the flip side, namely how challenging it is to build scalable products in an organization that had previously relied on project work and research projects for funding. However, the positive aspects far outweighed the challenges, and I wouldn't want to miss that time for anything. Even today, I am friends with my former colleagues, and I am immensely grateful.

For me it was the perfect entry into the world of technology, because I was able to grow every day through the challenges. I was constantly given tasks that were completely new to me, but through the trust placed in me, I not only developed my knowledge rapidly but also gained much more self-confidence. I learned that, as a woman with a business background, I can delve into technological challenges and provide significant value.

This realization subsequently helped me apply for a job that actually required a technical background and allowed me to secure the position. This is a valuable lesson I can pass on to other women: the job description or job offer doesn't always have to align perfectly with your skills.

In fact, I believe that if a job description already matches you 100%, you should consider whether you really want the job. If you already possess all the qualifications at the start, where can you further develop yourself? My first learning experience was that a job must also align with a person's personality. My second learning builds upon that, emphasizing the fact that you must have confidence in yourself and trust that you can tackle new challenges.

Is it easy? No.

Does it require stepping out of your comfort zone? Yes.

Is it achievable? Absolutely.

11.4   MY ENTRY INTO THE WORLD OF STARTUPS, MOBILITY, AND BACK TO A LARGE COMPANY!

I had been interested in the startup world for quite some time, and fundamentally, entrepreneurship and the fascination with complex organizations, which companies inherently are, were the reasons why I studied business administration. I even had considered starting my own venture one day, but I always felt that I needed to learn more before I could do so. Additionally, I had also realized that I needed variety in my workday.

So, it seemed like an exciting challenge to be part of building a startup initiative that aimed to connect young startups with an established company in the automotive industry. I had already learned a lot about innovation in business modeling and had experienced how difficult it can be to successfully drive innovation. However, I had little understanding of what it meant to foster external innovations from startups within an already highly innovative established company.

When I started, our team consisted of three people, including my boss, but I was the only one located at the headquarters in Graz. At that time, my boss and my colleague, who had developed the concept for the corporate startup initiative, were both based in Germany. Therefore, I had to quickly familiarize myself with the tasks and also understand how the processes of this company with over 10,000 employees worked. At the same time, I had to help establish processes for the newborn startup initiative. Furthermore, I found myself in a company that had existed for over 70 years, known for its innovations in the automotive world, and accordingly had an incredibly vast portfolio.

In the beginning, many things were difficult for me to grasp, and honestly, I still discover new things today. But the atmosphere was very good, and I was fortunate that many people took the time to introduce me to their areas of expertise. Gradually, I began to understand more, and every day I learned about at least one new startup in the mobility field. Suddenly, there was so much variety that I wasn't sure if it was possible to proceed in a structured manner.

I had doubts at times about whether I could handle it, but at the same time, I was also given a lot of trust, and my boss repeatedly threw me into the deep end. I participated in tech meetings on short notice, presented the initiative and the company at events and even in front of the camera, and in the first week, I was assigned my first startup project to take care of. Having gained a lot of trust, but also with it responsibility and a lot of freedom in a short time, I started to work creatively. I realized that I not only had the chance to contribute to building something, but I also had the opportunity to work creatively and make a difference.

Furthermore, through the many startups I interacted with every day, I gained deep insights into new technologies and market trends, and witnessed an incredible variety of business models. The first year was probably the most intense of my career so far, but it was also the year in which I learned the most. It wasn't just about immersing myself in the technology; it was also about developing the understanding necessary to bring together two such contrasting parties as a startup and an established company.

Our goal was and still is to co-innovate with startups, and in doing so, one thing plays a significant role: empathy and the ability to take people on this journey. Startups tend to have a clear focus, rapidly prototype, and test things in the market as quickly as possible. This approach helps them discover and build their customer segment, learn about their pain points, and develop solutions that closely align with their needs. Trial and error should be the order of the day to develop successful, resource-efficient solutions.

On the other hand, an established company has already developed its business model, typically knows its customer segment, and aims to satisfy it with 100% solutions. New developments or changes require approvals, follow processes, and must adhere to standards and quality criteria that allow larger companies to run their ongoing business successfully. That's why it tends to be more challenging for larger companies to enter new areas or market segments. Startups can move faster and be more agile in this regard. Therefore, one can imagine that two different worlds collide here, and of course, each approach has its merits.

Nevertheless, for a co-innovation project, a common approach must be found, and often, a mediator is needed. It was precisely this mediator role that helped me to grow personally and become more resilient. And that's where I believe the third lesson lies, which I would like to share: resilience. I believe that becoming more resilient is a lifelong process, but I have certainly learned that it is important to train this ability to be able to bring about change and stay fit in the process.

I think the following quote from Maddi and Khoshaba, who conducted a major study on workplace resilience in the United States, describes it very well:

> Simply put, these attitudes are commitment, control, and challenge. As times get tough, if you hold these attitudes, you'll believe that it is best to stay involved with the people and events around you (commitment) rather than to pull out, to keep trying to influence the outcomes in which you are involved (control) rather than give up, and to try and discover how you can grow through the stress (challenge) rather than bemoan your fate.
> (Maddi & Khoshaba, 2006)

And I believe that this ability, like a muscle, needs to be constantly trained, especially if we want to shape the future!

11.5   WOMEN IN STEM AS DRIVERS OF INNOVATION

Innovation, to me, means the future! I see it as a tool with which we can actively shape our future. That's exactly what fascinates me, and that's why I work in the field of innovation and STEM. I love the potential that technology gives us to change things, and even more, I believe in the positive change that technology can enable! I don't believe that everything was better in the past, and we should orient ourselves towards it. Is every new invention good? No. Did everything go right in the past? No. Will everything go right in the future? No.

However, do we have the potential to steer things in a positive direction? Absolutely, yes!

I don't want to claim that women can do everything better or would have done everything better, but I believe that diversity contributes significantly to successfully shaping our future. Let me briefly justify my thoughts using the definition of innovation.

There is no universally accepted definition, but many are influenced by Schumpeter, who coined the term through his innovation theory, describing it as the introduction of technical or organizational novelties. He laid the foundation for recognizing ideas, inventions, and innovations as distinct concepts, stating that an innovation can only be declared as such when it is also accepted by the market. In the United States, innovation is often described as "knowledge making," emphasizing the connection between innovation management and organizational learning (McElroy, 2002).

I personally agree with both definitions, but I often find that the human aspect is missing, which, in my opinion, should be at the center. Ultimately, innovations are driven and used by people. Innovations primarily emerge at the interfaces between systems and cultures and through dialogue with various stakeholders (Rogers, 1983). Eventually, most companies would agree with me that their innovation success is ultimately achieved by employees who develop, refine, and implement new ideas, and that it is important to keep the user in focus to bring successful solutions to the market.

Let's take the mobility sector as an example. It not only undergoes many changes but has also been heavily male-dominated in the past, resulting in services tailored to male needs. A well-known example is crash tests conducted with dummy dolls that exclusively represented male physiques in the past, leading to more fatalities among female accident victims.

Let's think further and not just talk about the lack of influence of women. Let's consider other countries that often have different mobility needs than highly developed countries. These countries are largely excluded from the development process, despite efforts to expand the market to these regions. However, it is important to emphasize that I do not believe this is a malicious intent in product development. It is simply challenging to develop solutions for diverse groups (and our global population is indeed diverse with various needs) when they are not or only minimally involved in companies and the development process.

Diversity and innovation go hand in hand and are essential elements to succeed, especially in complex times like the ones we are experiencing now!

Diversity in innovation is crucial for several reasons. Firstly, it enhances problem solving by bringing together individuals with different backgrounds, experiences, perspectives, and knowledge. This diversity of thoughts enables a wider range of problem-solving approaches and solutions. Secondly, diversity fosters creativity by encouraging the exploration of different viewpoints and ideas, what leads to fresh and innovative concepts. Additionally, a diverse workforce helps organizations understand and cater to a broader range of customers and markets, expanding their market reach.

By avoiding groupthink through diversity, organizations promote constructive debate and critical thinking, which leads to more innovative outcomes. Groupthink typically occurs when a homogeneous group of individuals uncritically agrees with each other, and this clearly blocks innovation.

Moreover, diverse teams make better decisions by considering a wider range of perspectives and avoiding biases. Research has shown us that diverse groups tend to outperform homogeneous groups in terms of decision-making accuracy and effectiveness. Lastly, diverse teams are more likely to identify a broader range of problems and opportunities, tapping into a wider pool of insights critical for innovation.

So for me, innovators are individuals who identify problems and seek new solutions. It is a characteristic of people who open their eyes and want to tackle problems. Personally, I believe that innovation should aim to improve our planet and our lives, and therefore, it should address ecological and social issues.

I've spent a lot of time over the past year trying to understand how we can pivot innovation in a more sustainable direction. Basically, what I'm trying to do here is bring together all my previous experiences because I firmly believe that we need to rethink business models in order to develop real leverage for sustainability while maintaining an entrepreneurial and financially successful approach. Technology plays an essential role in this and opens up new possibilities for us. When used and understood correctly, it can effectively address pressing social and environmental issues. I have started giving workshops and seminars on creating social and planet-centered innovation because I have a strong belief in the importance of this topic.

To achieve that, we cannot think only about individuals; we must think and act globally, and that requires diversity within companies.

ABOUT THE AUTHOR

Viktoria Ilger is a dynamic business economist whose great passion is new technologies and their potential for people and the environment. After spending some time in the banking world, she dedicated herself to innovation. She started her innovation career in a small software company, where she worked closely with clients on digital business models. Over the last five years, she has been instrumental in building the AVL Startup Initiative and has led the Startup Innovation team since 2020. Viktoria is a visionary and in recent years has increasingly focused on how innovation and sustainability can go hand in hand to ensure the future viability of our planet and companies.

REFERENCES

Maddi, S. R., & Khoshaba, D. M. (2006). *Resilience at work: How to succeed no matter what life throws at you.* New York: Amacom.

McElroy, M. W. (2002). *The new knowledge management: Complexity, learning, and sustainable innovation.* Boston: Butterworth-Heinemann, USA.

Rogers, E. M. (1983). *Diffusion of innovations.* 3rd ed. New York, London: Free Press.

12 Safety Boots to Break Glass Ceilings™

Emily Soloby
Juno Jones®, USA

Why do STEM and trades fields have such a hard time attracting and keeping women on board? The feeling of invisibility, of not being acknowledged, and of being overlooked or outright discriminated against runs deep. One part of fighting the culture that causes this is making sure companies are supplying women with the proper PPE (personal protective equipment) and safety footwear. As advocates, we cannot do this without raising awareness of the issues, and companies cannot solve the problem without providing safety footwear options. That is where Juno Jones comes in. But my own journey begins a bit farther back in time.

12.1 PASSIONS COLLIDE

I probably remember every pair of shoes I've ever owned. There were the 1980s pastel Reebok and white leather Nike high-tops, the black combat boots, the saddle shoes from Journeys, the white brogues from Paris (from an epic family trip when I was 14), the motorcycle boots from Zipperhead on South Street in Philly, the Italian knee-high go-go boots from Trash and Vaudeville, my first platforms (suede sandals from Oxford Street), the rugged brown Chelseas that took me across Ireland during my semester abroad (purchased in Georgetown, Washington, D.C.).

I've always loved shoes, but not in the Imelda Marcos way of collecting hundreds. I've loved shoes in the sense that they have always been something carefully selected, something that brings me back to a moment in time, something that defines me, and stays with me forever, if not in a physical sense, then in my deepest memories. Perhaps it is part of my DNA. A portrait hangs over my mantle today of my great-great grandfather Salvatore Rubino, who was a shoemaker in Sicily. I have had dozens of pairs of shoes that, in my mind, I will take to the grave.

Not to put too serious a point on it, but as you can see, I feel strongly about shoes, and how, by virtue of helping to define our style and by virtue of accompanying us on our life's journey, they truly become a part of who we are, who we have been, and who we can become.

Although I love shoes and, thanks to my stylish mother and grandmother, fashion in general, I never thought of it as a career. Social justice has always been my passion, specifically women's rights. I can remember in elementary school, hearing the phase passed down from my French great grandmother that it hurts to be beautiful (*il faut souffrir pour etre belle*), and thinking, why? Why should women suffer in

DOI: 10.1201/9781032679518-14

order to be beautiful? Rejecting that notion was one of my earliest acts of rebellion against a system that expects us to suffer for the pleasure of men. That will never be me, I thought.

But still, I was the child who wore a frilly dress on the weekend simply because I loved it. Somehow at that early age I knew, just like the popular hashtag says on Instagram today, that #youcanbeboth. You can be both strong and girly, you can be both skilled and stylish, you can be both powerful and feminine. I had powerful role models to admire in my mother, who easily could have been a beauty queen, but who instead was the first woman to train with the Rutgers University track team (she was only in high school at the time), and my grandmother, who encouraged my education and always warned me "never to clean up after boys."

I spent my high school and college years working on social justice issues, attending demonstrations for women's rights in Washington, D.C., heading up clubs and organizations, and advocating for women whenever I could. I was a women's studies major, and at the University of Minnesota I got an internship as a domestic violence courtroom advocate, where I learned the legal system around these issues and advised women in court as they stood up against their abusers. During these years, I also took my first shoemaking course, just for fun, in Cuernavaca, Mexico, during my study abroad program. When I got back to the United States, it was almost time to graduate, and I decided to continue with my women's advocacy work via the legal system.

During law school, I worked as a legal intern at the National Organization for Women in Washington, D.C., and once I graduated from law school and passed the bar, I went to work as an attorney for Legal Aid, helping women and children in equitable distribution and custody matters. Working as a legal services attorney is not for the faint of heart. After a few years, I knew I had helped many people—I had the handwritten letters and thank-you crafts to prove it! I held these dear, but I knew it was time to move on.

Over the years, whenever I needed a job or a break from what I was doing, I was very lucky in that the company my father worked for was often able to take me on part time. It was a chain of fine jewelry stores, and it was there that I learned to connect with people while helping them to find the perfect birthday gift, engagement ring, or graduation watch. I learned this through watching my dad, who is an amazing salesman. The key to being an excellent salesperson, I learned, is simply finding a way to connect with the customer, and truly wanting to help make them happy. It's not about "making a sale" for yourself. Instead, it's about providing information and helping people make a great choice for them.

After my time at legal aid, I returned for a little while to the safe haven that was the jewelry store. There, I laughed with my customers and coworkers, walked the mall studying my favorite thing (shoes), and thought about my next move. I knew I still wanted to make a difference in the lives of women, and I thought a way to do it would be by listening to and sharing their stories. I decided to return to school for my master's degree in broadcasting, telecommunications, and mass media. It was a competitive program with world-renowned teachers, and I was in my element discussing the effects of media on our culture with other passionate students. One of those passionate students was named Ryan, and I didn't realize it at the time, but he would become my husband and the love of my life.

After completing the two-year program, Ryan and I graduated with our master's degrees and were working in the field. But then, an interesting offer came in from Ryan's uncle, who had started a small truck driving school in Harrisburg, Pennsylvania. Uncle Mike wanted to retire soon, and offered to teach us the business so that we could eventually buy it from him. Being entrepreneurial and adventurous (and not having any kids yet), we decided to go for it. We each spent time in Harrisburg learning the ins and outs of that business, and Ryan got his CDL and tester's license. Together, the two of us went before the Pennsylvania Board of Education and received a license to open a new school in the Philadelphia area as well. Those early years at AAA School of Trucking were filled with long hours and encompassed everything from high-level meetings to taking out the trash. We even took our new baby to the office with us and let him sleep in the Pack and Play while we worked at our desks.

12.2   THE SPARK

It was during those years that I began to notice that women's personal protective equipment, or PPE, for the jobsite was sorely lacking. What was out there was "unisex", which really means men's, because it's always made around the basis of a man's body. It was gigantic, it lacked the proper shape and fit, and—let's face it—it looked bad. Of course, my main area of concern was shoes.

I'm the kind of person who likes to have the right outfit for the occasion and the right equipment for the job. I owned a pair of traditional work boots, but I just didn't want to wear them. When I put them on with my jeans and blazers, they gave me a clunky clown look with their bulbous toes and masculine appearance.

Someone recently asked me what it was like working in the male-populated industries of trucking and footwear. The way she phrased the question with these two separate parts of my career made me think about the differences. The trucking school is a business that my husband and I own and run together, so the dynamic is so much different. When we show up to a meeting with people we don't know, they may see us and automatically think, "Oh, it's the owner and his wife." I've had people ask me if I answer the phones. Even if they are keeping it to themselves, the assumption in those early days was often that this truck-driving-related business belonged to my husband, and that I was the helper.

In meetings, when I spoke to an unfamiliar group, they may have listened to me a bit, but then looked at my husband to see what he had to say. They may have interrupted me, or talked over top of me. Unfortunately, this is the plight of many women in male-populated fields. However, at Juno Jones, I am the Founder and CEO. Even though at this business, I still work with my husband, who is a partner in the business, I am known through my business presence as the face and voice of the company. So, as you can imagine, I'm perceived differently. In a meeting of unfamiliar people, when I speak, something different happens: people quietly listen to what I have to say, nod, agree with me, address me directly. The contrast is palpable.

So, when I was building AAA School of Trucking with Ryan, running from my office to job sites and to client meetings, it was extremely important to me that I be taken seriously as an equal at the company. I wanted to be seen as the leader of the organization that I was. When it came time for me to put on my ugly tan workboots with my sleek jeans and blazer to go out to the work site, I didn't want to do it. I knew the clunky "too-big safety boots" look was only undermining my credibility. I remember going back to the office after one of those meetings, and immediately searching Google for "stylish women's safety boots." The year was 2018. What do you think I found? Nothing. I kept searching, and I saw only more of the same.

I was sitting at my desk thinking about this when my extreme annoyance began to take shape into something else—an idea. I had always wanted to make shoes and dreamed of one day starting my own shoe company. Back at my house were boxes of cut-up shoes cobbled together into new shapes and designs. The leather sandals my sister wore every day were ones I had made by hand. If there was a gap in the market for women's safety boots that needed to be filled, why couldn't I do it myself?

I remember the moment I told Ryan about the idea. We were walking up State Street in Chicago ahead of the wedding of our friends Shayna and Larissa. We had all waited a decade for this amazing wedding to be legal, and our energy and excitement was high. Ryan is always supportive, but at the same time honest, so I knew if he thought the idea wouldn't work, he'd tell me. In fact, we often have some level of friendly disagreement over big decisions which leads me to advocate for my positions, so I was a bit nervous to see what he'd think.

"Are you ready?" I asked. "I'm really excited about this." And then I spilled it. Stylish safety boots for women. Steel toe boots that women would actually want to wear and that would fit women's feet properly. Silence. We walked a few more strides. "Yeah" he finally said. "I like it." Nothing better than having your partner on board with your zany ideas!

I knew I had finally found my own way to advocate for women in the trucking industry, and soon I realized I had the ability to expand that into empowering women in male-populated and hazardous industries across the board.

12.3   A SIGN

As an admirer of the brand Uggs, I was in awe of the reach they had accomplished, and I knew it was all due to entrepreneur Brian Smith, who took the brand into the United States and made it what it is today. I wanted to buy his book, but every cent mattered in keeping costs down. I scoured eBay and found a used copy for a few dollars. The day it arrived, I opened it excitedly, eager to devour every page. I cracked the front cover, and a wave of shock washed over me. I hadn't just bought an inexpensive used copy of "The Birth of a Brand." It was a rare signed book by Brian Smith himself! Could this be a message from the universe? I launched into the first chapter, which was about … Brian's own signs from beyond.

12.4   THE GROUNDWORK

Back in Philadelphia at the truck driving school, the idea continued to brew. As I was a former attorney, the first thing that concerned me was trademark. I set my sights on finding the perfect name and URL, creating an amazing logo, and getting the copyright secured. As my now two children played on the floor at my feet,

I scoured the web for name inspiration, read a PDF called "The IGOR Guide to Naming" and a book called "Hello, My Name is Awesome." I wanted to choose a name that represented a powerful, free, modern woman, who wouldn't balk from trying new things and, yes, spending her days as a badass in a male-populated industry. In my mind, she was a real woman, and at the same time, a superhero.

But as a lawyer, I knew there was a lot more to finding the right name than just coming up with it. It had to be available, as in, I had to search the TESS System to make sure no other apparel, footwear, or similar companies had already laid claim to a similar name or had used it in commerce. Fun fact: the original name of Juno Jones was actually "The Adventures of Juno Jones," for trademark reasons, but we were eventually able to shorten it and secure the needed trademarks for the name Juno Jones™ on its own. The URL was another story—www.junojones.com was taken at the time, and it seemed to be partially in use by a grocery market. We settled at that time instead for www.junojonesshoes.com.

A year or so later, I applied for Juno Jones to be a part of the Philadelphia Fashion Incubator at Macy's. This was a nerve-wracking process filled with forms, essays, portfolios, interviews, and presentations. I had gotten through the first round of the process when it came time for the board to review our website. Instead of typing in www.junojonesshoes.com, they assumed it was the shorter version and just typed in www.junojones.com. I got an immediate alarmed call from the director, letting me know that "your URL is leading to a porn site." You can imagine my horror!

I assured them that was not our URL, but that we were working on getting it for ourselves. At the time, it was thousands of dollars and out of our reach. The fashion incubator board got over the shock, and thankfully, Juno Jones was chosen for the spot. For the next couple of years, I babysat it using GoDaddy's service, and one day, in 2021, I received an email—the porn site had let it expire! We quickly snatched it up. Whew! It felt great to know that www.junojones.com would now be ours, and would be now used to empower women and keep them safe.

12.5   THE DESIGN

The trucking industry is my "home" in many ways, so when it came time to get feedback about Juno Jones, the first place I went was the Women in Trucking Association. Founder Ellen Voie was enthusiastically supportive of the idea for Juno Jones from the start. She helped me connect with women across the United States, and even Canada, to learn about their needs in safe footwear in transportation. It was there that I began to gather data for what types of safety features were needed, and what materials, colors, and styles women preferred. After I gathered this information, I began to branch my research out into other industries as well. On Instagram I met so many inspiring women working in industries from manufacturing, to construction, to engineering, and more.

I had studied shoemaking and made shoes by hand at the Brooklyn Shoe Space in New York, and I had my own drawings, but I knew that with the help of a technical designer, we could take it to the next level. I was thrilled to work closely with Amanda, who had previously designed for such classic brands as Cole Haan, Ann Taylor Loft, and Naturalizer. I knew that she understood both timeless style and comfort. As we listened carefully to what women were telling us in chats, phone calls, surveys, and in-person focus groups, we formulated sketches and gradually narrowed them down to a few styles. Although there were so many boots I was dying to create, I knew that for practical reasons, we'd have to choose just one to start with.

Amanda and I went through dozens of style sketches—knee-high boots, booties that barely grazed the ankle, chelsea boots, lace-up boots, engineer boots with buckles, and of course, the boot that eventually became our hero style—the zip-up *Jodhpur*, a.k.a., the *Meti Boot*. This boot was the number-one favorite among the women we talked to for many reasons. First of all, it was a classic style, evocative of classic English paddock boots, and second, the comfort and ease of use was unbeatable. They went easily from the office to the job site, and looked amazing with jeans, cargo pants, slacks, and dresses alike. We knew we had a winner in this style, and couldn't wait to get started with development. But, how would we find a manufacturing partner?

12.6 MANUFACTURING

Our search began in the United States, as it would have been our ideal place to produce. Unfortunately, the infrastructure for footwear manufacturing left US shores long ago. While there are still some US-based factories, most of them are owned by existing large companies who make only their own products. The others are making only boots on a very small scale, already have enough clients (think US military) or are not experienced with safety boots.

We knew we would need to take our production out of the country, but where? Our search led us first to Europe, as Italian and Portuguese shoes are famously gorgeous and well made. Plus, what better place to visit on a regular basis than a stunning European city? Unfortunately, the cost was more than we wanted to have to charge our customers, who we knew needed these shoes for work. I thought about where some of my own favorite classic, durable, and rugged boots were made—Mexico, of course!

We were working with a consultant who recommended against Mexico as it was "too dangerous." I had spent a lot of time in Mexico, having studied Spanish and shoe-making in Cuernavaca as a college student, and returning later as well, so to hear that it was too dangerous seemed inapplicable. We parted ways with our consultant and headed down to León. It was there that we met with over a dozen factories to explain our mission, and the types of boots we wanted to create. Between my basic Spanish and their excellent English, we were able to understand each other perfectly.

Ryan and I spent two full days touring factories and meeting with makers and suppliers. At the end of our second day, we had amassed a list of people to follow up with. My brain swirling with Spanish, I was sitting in the waiting room of our last factory of the trip, and thinking about arriving back at the hotel, eating a giant meal, and falling asleep. A woman popped her head into the waiting room and said they'd be with us soon. I watched the clock on the wall ticking, and tried not to close my eyes.

Suddenly, the door burst open and a young guy stood there, beckoning us in. He invited us to sit, and my fatigue vanished. His energy filled the room, and as we explained our brand and what we wanted to accomplish, he got more and more excited. This was a fourth-generation boot maker, and he and his brothers were the new generation. They had energy, they had growth in mind, and they "got" us and wanted to be a part of it. I knew at that moment that Juno Jones had found its perfect partners.

12.7   THE LAUNCH

With our production team lined up, it was time to get ready to launch our Kickstarter. Kickstarter is a platform that allows "backers", or customers, to pledge a pre-purchase of a new product. The product cannot already exist on the market, and there is no guarantee to backers that the product will ever get made, or that they will ever receive anything for their money. In addition to that caveat, if the creator doesn't reach their set goal, they don't receive any of the money at all from the campaign, and backers get an automatic refund.

Backers are putting a lot of faith in and betting on the creators, and creators are staking their time and financial investment in creating the Kickstarter campaign on the hope that people will back them. With that in mind, we knew we had to spread the word about Juno Jones well before we launched. Without any investment funds, it was up to us to raise awareness and form a community around the brand prior to launch. Every night after my family was asleep, I would sit on my phone, scouring Instagram for women who I thought would love Juno Jones. I'd reach out to them and tell them about the idea and the brand.

We chose February 11, 2020, the International Day of Women and Girls in Science, for our launch. As the day neared, we got more and more excited. Our small team was doing everything we could to promote the launch through our email list, Instagram, and Facebook. Our factory would be ready to go with production as soon as they got the word, and all systems were go. After working so hard up until this point, it seemed surreal to be about to hit that "launch campaign" button on the Kickstarter page. This had been years in the works. I watched the clock and when it hit the designed time, I clicked.

I quickly navigated to the Juno Jones social and Mailchimp and let everyone know—the campaign was LIVE! We got our first backer. It was one of the amazing founders of the group iFundWomen. They didn't need boots, and we had not been working with them—they were just cheering us along. Soon, there was another ping. Then another! Someone pre-ordered two pairs. Someone pre-ordered five pairs for her team at work! Then, a miracle happened. Kickstarter took note of the early support we were getting, and listed us on their home page as a "Staff Pick". Within 29 hours, we had funded our campaign 100%. The *Juno Jones Meti Boot* was happening.

12.8   LOYALTY

Little did we know the challenges that were right around the corner. We had heard about the "Coronavirus", of course. It was an unsettling piece of news coming from across the world. Every night, lying in bed, thoughts of Covid-19 popped into my head as I tried to focus on sleep. I made a conscious decision to push them away, to pretend it wasn't happening. But people were dying. And they were dying closer and closer to home. I tried to focus on Juno Jones to fall asleep. It was mid-March when Covid shutdowns began in the United States. Our children's schools shut down, and Ryan and I were mandated by the city to temporarily close our trucking school as well.

We set up shop in our house, Ryan creating a makeshift office on the floor, using the kids' toybox as his desk. We patiently awaited news from our factory on the production status, but we knew that what would happen in Mexico was the same thing that had happened in the United States. Our factory was forced to temporarily close for safety from Covid-19. We fully agreed that this had to happen to keep everyone safe and to minimize the spread of the virus, but we were not sure if our customers, who would already be waiting months for their boots, would be as patient and understanding. They were.

As it became apparent that boot production would now take much longer than originally promised, and we shared this with our customers, we found out that they were not just customers, but true supporters and friends as well. Nobody got angry. Nobody grew impatient. Every one of our backers stayed the course, waited patiently for the boots they had ordered, and remained steadfast and kind. For this we will be eternally grateful to each and every one of them.

12.9 HAZARD GIRLS

But what could we do during this year of waiting? It was on a flight back from Leon that I had come up with the idea for "Hazard Girls". I had so many chats going in Instagram DMs that it was hard to keep track. So many of these women were craving conversation around the issues, and several people asked me if I could introduce them to others in their field. One woman told me she was an engineer in a small town and did not know one single other woman in a similar field. I had been active in several Facebook groups at that time for various fields, but I realized that there was not really a group that combined all male-populated fields under one umbrella. We put together the Hazard Girls Facebook group in an afternoon, and by the next day, it was filling up.

The Hazard Girls Facebook group was certainly my lifeline during this difficult time of wiping down groceries, home schooling a kindergartener and a third grader, and generally worrying about my family and the future. I loved that I had the group to go to for commiseration, discussion, and camaraderie.

Some of my hours during lockdown were spent as a guest on different podcasts. One of these, in fact the very first one, was the WAM, or Women and Manufacturing Podcast. I was interviewed by host Linda Rigano, and as it was my first time, I was just a little bit nervous. It's hard for me to believe this now, but I actually scripted out the entire interview, and when I was asked questions, I read the answers from my script.

Apparently, that master's degree in broadcasting paid off, because soon after my interview aired, I was offered a position as host of that very podcast. I hosted an episode or two, and before I knew it, the producers called me and asked me if I'd like to launch our own podcast, called Hazard Girls (Women in Nontraditional Fields).

It was an exciting offer, and I jumped on it immediately. I imagined being able not just to help women by bringing them options in safety footwear, but by telling women's stories as well.

I was thrilled, and we dove right in, creating a new episode every week. Some of the amazing women we have had the chance to interview include Mitzi Perdue (rice farmer, agricultural entrepreneur, and widow of Frank Perdue), Allison Grealis (head of Women in Manufacturing), Anne Pfleger (president of NAWIC at the time), and Katya Echazaretta (the first Mexican-born woman to go to space—there is even a Barbie version of her!).

12.10   HOLDING OUR BREATH

As a world, we collectively waited for Covid to "end", and as a community, the Hazard Girls collectively waited for Juno Jones boots to be made. Finally, one day, our backers got the news—the *Juno Jones Meti Boot* was on its way! We posted diagrammed maps showing the moves of our truckload of boots traveling across the border from Mexico in to our warehouse in Pennsylvania. They arrived at our warehouse in one piece, and our staff began shipping them out immediately.

Every order got an envelope of limited-edition stickers and a handwritten note from yours truly. (Yes, my hand almost fell off, but it was worth it.) The moments from the time the *Meti Boots* shipped to the time they began arriving at backers' homes was one of the most excruciating of my life. Would people love the boots as much as we hoped they would? Would they fit? Would our supporters feel that they were worth the wait??

I was frantically scrolling Instagram when I saw the first post. "They've arrived!!" I screamed to Ryan and the kids. We were all still working from home. The first review was from a stylish and talented architect I had seen on Instagram "finally—a woman's work boot that works!" she said, and posted several pictures. Throughout that day and the next week, the reviews started rolling in. "Worth the wait!", "So beautiful that I don't want to mess them up!", "They're so nice my husband doesn't believe they're work boots!", "I'm so happy!" The praise went on and on, and we couldn't have been more thrilled.

12.11   CUSTOMER STORIES

One of the most gratifying aspects of creating Juno Jones has been hearing the stories of women who have benefitted from our hard work, and who have found empowerment and joy in our boots.

When we were preparing the Kickstarter campaign, I met a woman, Adrienne,[1] who told me a story about finding out with a few days' notice that she would need to be on a construction site for her STEM job. She went shopping, but couldn't find any safety boots that would fit her small feet. When she arrived at the facility, they would not let her enter without the proper PPE. As a woman in a male-populated industry, she was a minority, and she faced the issues that as women, many of us face in these industries.

She was often undermined, made to feel like an outsider, made to feel "younger" than she was by being treated in a childlike way, ignored, and even sometimes harassed. The fact that she did not have the proper PPE to enter the site made her imposter syndrome kick in, as if she was somehow messing things up again by virtue of her very identity. So she was a bit relieved when the staff told her they had some loaners and went to fetch her a pair of steel toe boots. They brought her a typical pair of men's work boots in the smallest size they had. Reluctantly, she put them on.

Her heart sank as she went from feeling bad to worse. Not only did she not have the right equipment for the job, but now she was walking around feeling like, in her words, she was wearing her "daddy's safety boots." If she had felt delegitimized before, she now felt ridiculous as well. Any woman on the job will tell you that without the proper equipment, not only are you putting your safety at risk, but your very dignity as well. Needless to say, she was so happy to receive her stylish Juno Jones boots that fit, were comfortable, and allowed her to look and feel in charge.

Another customer, "Stormwater Nerd" Nadean Carson, found us via the National Association of Women in Construction. She is a civil engineer who worked in various industries over the years, then found her niche in construction stormwater management. After seeing contractors not have the labor or knowledge to correct the items she identified during inspections, she decided that she had to start her own construction business.

While dreaming and planning to start this venture, she bought a pair of Juno Jones boots, slipping them in the back of the closet until it was time. The culmination of a business incubator program that she had participated in was a business pitch presentation to potential investors. She knew that was the day to break out her new boots. She bought her Juno Jones boots while daydreaming about her company, and then wore them to debut Oya Construction to a group of over 20 community leaders. These boots solidified her venture from, in her words, "company cog" to business owner, giving her confidence and support as she began down the next exciting road.

The day finally came when it was time for her to present her business before the licensing board. Instead of wearing dress shoes, or ugly work boots that didn't fit, she told me that she donned her Juno Jones steel toes, and marched into the meeting with her head held high, knowing that she deserved to be not just a part of the industry, but a leader in it. She felt that she not only belonged, but needed to be there, and had the boots to prove it. She was granted her license, and today, her business is thriving.

You might have heard of our customer Karen Laine through her HGTV show *Good Bones*. Karen and her daughter buy and renovate homes to sell. Their brand of mother-daughter teamwork, business savvy, humor, and style is utterly addictive to watch. When Karen found out about Juno Jones, she was so excited because as someone constantly on demolition sites, nails are a major hazard for her. She told us a story about searching and searching for women's safety boots that fit.

Unfortunately, it was before Juno Jones had launched and could not find anything. One day on the work site, her fear came true, and wearing her not-so-safe safety boots, she stepped directly on a nail which went right through the sole of her foot. When a few years later she discovered Juno Jones with our ASTM puncture-resistant midsoles, she was thrilled to know that this type of injury would never happen to her again. Karen even made a video all about it to share with her audience.

Filmmakers love Juno Jones boots because they are often around heavy and rolling equipment, and on construction sites and in muddy, uneven terrain, so protecting their feet is paramount. Sarah Bullion heard about Juno Jones through a Facebook group for women in the film industry. She is an award-winning film director and part of the prestigious Alliance for Women Directors. Sarah was a Kickstarter backer and supporter from the start, promoting us within the film industry.

When Sarah was diagnosed with breast cancer, it was a major blow, and her friends rallied around her as she went through her treatment and recovery. One of the most beautiful posts I've ever seen on social media was the one she posted after finally returning to a movie premiere after this devastating struggle through chemotherapy. She stood tall and glamorous, and chose to wear her Raven Juno Jones boots to symbolize her own strength, power, and resilience. She's a true warrior, and nothing has ever made me prouder than seeing Sarah shining on the red carpet in her Juno Jones.

12.12   GRATITUDE

For me as Founder, the Juno Jones journey has been terrifying, sleep depriving, harrowing, and the most exciting and gratifying adventure I've ever embarked on. As a mission-based entrepreneur, it's easy not to lose sight of my goal, because it is a clear goal designed to do what I've wanted to do for my whole life—empower women. But one thing I never planned for or expected was some of the exciting things that have popped up along the way.

Yes, I have always loved fashion, but never did I imagine or dream I would present at the opening night of Philadelphia Fashion Week. Never did I imagine I would walk the runway with my models and wave at the cheering audience. Never did I anticipate that my work would be featured in art galleries, and as part of the Stuart Weitzman exhibit at the Michener Museum of Art.

We never hired a PR agent, and so never dreamed that we would be featured in over 60 media publications including the top footwear industry magazines, in international fashion news, and on TV. Never in my wildest dreams did I realize that I'd be given awards for my work. I am immensely grateful for these opportunities and honors, and to the friends and partners I have made along the way.

Some of the many partners we have made include the National Association of Women in Construction, Women in Trucking, the National Society of Professional Surveyors, the Alliance of Women Directors, the Philadelphia Fashion Incubator at Macy's, Duluth Trading Company, Zappos, Reinvented Magazine, and the Princesses with Powertools Calendar, plus many more.

12.13   CONCLUSION

As an executive in the trucking industry, I faced an unspoken problem of bias in the workplace, but I knew it wasn't limited only to me. With a background in women's studies, women's advocacy law, and the transportation industry, I understood that the problem is a systemic one that needs to be addressed on many levels. As a lifelong boot lover, with experience in hand-making boots, my passions had prepared me for this challenge. In the fight for workplace justice, what better battle ground than safety footwear, and what better warrior than Juno Jones?

Having the proper-fitting footwear is crucial for preventing injuries, and as women, it also plays a huge role in our confidence on the job, as well as in our feelings of validity and legitimacy. Juno Jones will continue to fight the good fight, and our mission will remain steadfast. We will continue to empower women through safety boots, and in so doing, to make the workplace a friendlier and more attractive place to women, so that as women we will be encouraged to join STEAM and trades fields, we will be incentivized to stay in them, and we will have the tools to rise to the very top. I envision my own children coming of age in a world where women are not just welcome in every industry and every workplace, or celebrated as an example of diversity, but expected, accommodated, and appreciated as the norm.

I often get asked the question of my vision for Juno Jones—where I see us heading in the next five or ten years, and into the future. The goal for Juno Jones is simple but lofty at the same time—to become an internationally recognized brand, which provides women with options in safety footwear that actually fit their feet, are comfortable and look amazing, no matter what industry they are in. From the factory floor, to the construction site, to the busy professional kitchen, women are making great strides—and we want to be there with them every step of the way.

ABOUT THE AUTHOR

Emily Soloby, MA, JD, is Founder of Juno Jones, The Stylish Safety Boot Company™; Co-Owner of AAA School of Trucking, and Creator and Host of the top 5% ranked Hazard Girls Podcast and Community. She has been on a mission to help women from the beginning of her career. While earning her BA in Women's Studies, Emily worked as a domestic violence victim advocate. During law school she served as Legal Intern at the National Organization for Women in DC, which solidified her passion for advocacy work. Following law school, Emily worked as a trial attorney with Legal Aid, helping women and children in family law and domestic violence cases. She went on to receive her master's degree in broadcasting. As a lifelong boot lover, Emily spent time learning the craft of shoe-making in Cuernavaca, Mexico, and the Brooklyn Shoe Space in New York. In 2009, she and her husband Ryan took over AAA School of Trucking, a truck and heavy equipment safety training firm, which she has helped grow into a national business. It was during her years there that she noticed the footwear problem—there wasn't any! Determined to create safety boots that she'd actually want to wear, Emily assembled a team to bring Juno Jones to life.

Emily sits on committees with Women in Trucking, and Empowering Women in Industry, and is on the Board of Directors for NAWIC Philadelphia. Her work has been featured in art exhibits including the Stuart Weitzman "Walk This Way" exhibit at the Michener Art Museum. Emily spent 2020 and 2021 as a designer in residence at the Philadelphia Fashion Incubator at Macy's, and she's been featured in over 60 publications and other outlets, including *Women's Wear Daily*, *Footwear News*, and ABC-TV. Her awards include the 2021 Empowering Women in Industry Woman of the Year Title, 2022 Comcast Rise Award, 2023 Women'sNet Distinguished Business Winner, 2023 Women in Trucking Top Women to Watch in Transportation, and 2023 Visa She's Next in Fashion Award. Emily lives in Philadelphia, Pennsylvania, with her husband and business partner Ryan, their children, and a very loud hound dog. For fun she enjoys rooftop restaurants, karaoke, and relaxing with her family in the Pocono Mountains.

NOTE

1 Some names in this section have been changed.

13 Intersectionality
How to Transform Headwinds to Tailwinds

Erika Anderson [in Remembrance]
Georgia Pacific LLC, USA

13.1 WHAT IS INTERSECTIONALITY?

Intersectionality, a term first coined by esteemed black legal scholar Kimberle Crenshaw in 1989, is defined by the Oxford Dictionary as "the interconnected nature of social categorizations such as race, class, and gender, regarded as creating overlapping and interdependent systems of discrimination or disadvantage". Crenshaw describes it "as a metaphor for understanding the way multiple forms of inequality and disadvantage sometime compound themselves and create obstacles not understood within conventional ways of thinking about antiracism, or feminism, or whatever social justice advocacy structures we have".

Essentially, intersectionality is a way of acknowledging the various unique elements that comprise our social identities, the idea that they intersect and can influence how we experience, and thus navigate, the world around us. Granted, some aspects of one's identity may afford privilege, while others may result in oppression. We oftentimes primarily think about race, gender, sexual orientation, and/or class, but it should be noted that social identities extend to many other categories, such as age, language, ability, religion, nationality, and even educational background.

As a relatively young, disabled, queer, black woman working as a mechanical engineer in the "boys' club" that is the manufacturing industry, I can say that the road to both get here and stay here has certainly not been free of discrimination resulting from my social identity. Throughout my educational and corporate career journey, I've battled many headwinds tied to my overlapping social identities, which we now understand to be intersectionality. Yet, the best part of my story is I am still happily working in manufacturing and loving what I do, despite the headwinds.

As I share my story and the techniques I've used to transform headwinds into tailwinds, it is my hope that it will empower others from historically marginalized, intersectional groups to excel by applying these and comparable strategies. On the other hand, it is my hope that individuals belonging to historically privileged groups will better understand the oppressive impacts that one's social identity can have on social and professional mobility, and subsequently feel moved to become better intersectional allies, using their privilege to give voice and serve as tailwinds to marginalized groups.

DOI: 10.1201/9781032679518-15

13.2   THE COURAGE TO CHANGE THE THINGS THAT I CAN

Growing up, I was always considered a "gifted" student. Throughout most of my grade school years I took advanced classes and, once in high school, AP classes. As I got closer to senior year, I began to question whether I would really be prepared for college. Even though I always got all A's, my high school was considered a "bad" school, or at least that is what I always heard from other parents at other schools. Could it have been because we were in a low-income neighborhood where 98% of the student body was economically disadvantaged? Was it because 85% of the student body belonged to communities of color?

At that point I wasn't sure why my school was "bad"; I just knew that I wanted to be able to compete with my white counterparts in college, and quite frankly, I began to doubt if my high school could prepare me accordingly. Therefore, in an effort to get what I convinced myself would be a better education, I moved away from my home in Stone Mountain, Georgia, to live with my aunt in Germantown, Tennessee, a suburb of Memphis known for its top-tier schools. The young me just knew that if I could get A's at a school like that, then certainly I would be prepared for college. Unfortunately, this plan did not play out in real life the way I imagined.

On the first day of my junior year at my new school, I was more anxious than I had ever been on the first day of school. Not only was this my first time since elementary grades starting at a new school, but it was also in a new city and school district where I didn't know anyone. Additionally, this would be my first time at a school where a majority of the students did not look like me. Despite my angst, I was still excited about facing new academic challenges that my former school couldn't provide. For example, the list of AP classes offered was far longer than my old school. There were also more extracurricular activities beyond sports to explore. I wanted to immerse myself in it all and become the well-rounded academic student who all colleges would want on their campus. Yet, on the first day of school when sitting with a counselor to set up my class schedule, my hopes were crushed.

I already knew the classes that I wanted to take: AP English, AP US History, AP Physics, and Honors Pre-Calculus. It was important to me to take AP and advanced courses, as that is what stands out on college applications and that is what I was accustomed to. However, the counselor informed me that I would not be able to take any of the aforementioned AP courses nor Honors Pre-Calculus. I was stunned and speechless. My aunt, just as confused as I was, then asked why I would not be allowed to take those classes.

The counselor attempted to explain how this school is different from the education that I was previously receiving, more challenging if you will, and it would be even harder for "a student with your background" to be successful in AP courses at this kind of school. She tried to convince us that regular courses at their school would be the equivalent of the Honors courses taken at my old school and therefore, my 4.0 GPA was not the same as their 4.0 GPA.

Thus, to protect me from academic failure, it would be best for me to take regular classes. I insisted that I had the skills necessary to be successful and would be fine in those classes. However, she made it clear we had no choice in the matter, I was not allowed to take AP classes, but she was willing to compromise and allow me to take Honors English.

"Don't let her see you cry" is what I remember repeating to myself in my head. All the scenarios I dreamed up of me being in these different AP classes with diverse students, engaging in academic dialogue, and learning together, would never come to fruition. I was devastated. Never in my academic career had I been denied entry into advanced classes! I had only ever taken advanced classes. What would a gifted student like me do in a regular class? As I replayed her words over and over in my head, "a student with your background", it finally became clear what those parents meant by a "bad" school, when referring to my old high school. Their judgement had nothing to do with the actual students or our academic aptitude and achievements. Simply put, we were being judged by our economic status and the color of our skin.

After what felt like the longest first day of school ever, I rushed home to call my mom and vent about what happened. Naturally, she was sad to hear me in pain, yet her mom instincts still kicked in as they always did. She reminded me of a prayer we would say often in our home: The Serenity Prayer. *"God grant me the serenity, to accept the things I cannot change, the courage to change the things I can, and the wisdom to know the difference"*. I needed to look at this situation and understand what was in my control versus what was not. As unfair as it seemed, my class schedule was not in my control, I could not change it, and I needed to accept that.

Failure to accept this fact and dwell on the situation would only hurt me, no one else, by living rent free in my head, taking up valuable brain real estate better applied to my studies. What was in my control, however, was my academic performance in those classes. My mom also reminded me something I had clearly lost sight of: I was not smart because I took honors classes, I was smart because that is just who I am. I apply myself and I work hard. Classes do not define me. That counselor did not define me. Social stigmas attached to the type of school I attend did not define me. *I* define me.

I returned to school the next day more motivated than ever. This motivation carried me academically through the school year. I walked those foreign halls with my head held high, knowing that my skin color or economic status did not matter; I was smart, and I would prove it. Socially, however, I still struggled to connect with other students and began feeling lonely and isolated. Just when I thought I couldn't stand it any longer and wanted to move back home, my Honors English teacher, Mrs. Garrison, a compassionate middle-aged black woman, extended an offer to me to eat lunch in her classroom with her and her daughter, a senior at the school. I was so relieved and elated!

Finally, I didn't have to sit alone at lunch. Finally, I had another student to talk to. Finally, I felt like I belonged. By the end of the school year, not only had I made friends, but I made all A's. Knowing that I did not allow that counselor's judgement of my capabilities to dictate my performance was a very satisfying feeling. I was finally convinced that I would in fact be prepared for college, regardless of what high school I attended. I would be smart in any space because that is just who I am. Therefore, I went back home to finish my high school career in Stone Mountain with my friends and family, where I felt I truly belonged.

13.3   REFUSING TO LET ACADEMIC STIGMAS DEFINE ME

After many years of dreaming of going to college, I was accepted into Spelman College, an all-women's Historically Black College and University (HBCU). As a fifth-generation Spelmanite, I was proud to follow in the footsteps of women before me in my family. My love for math and physics in high school sparked my desire to study mechanical engineering (ME). Although Spelman does not offer an engineering degree, their distinguished Dual Degree Engineering Program presents students with the opportunity to spend three years at Spelman completing a STEM-related degree and then transfer to one of several partner engineering institutions for two years to complete an engineering degree. As an Atlanta native, I knew I would be transferring to the Georgia Institute of Technology, also known as the MIT of the South. Sure enough, after three years completing my bachelor's degree in math and taking pre-engineering courses, I was accepted into Georgia Tech's ME program.

Like my junior year in high school, I was returning to an academic space where I would be in the minority. In this case, not only was I one of the few black students in my classes, but there were also few women, and even fewer black women. Thankfully, I was at a point in my life where I was comfortable connecting and making friends with students from majority groups after having done several summer internships throughout college. Unfortunately, that did not prepare me to connect at Georgia Tech.

Engineering Dynamics proved to be my favorite course during my first semester at Georgia Tech. Not only was the professor very captivating during his lectures, but I had already taken this course at Clark Atlanta University; however, the credit did not transfer to Georgia Tech as I had hoped. Yet this proved to work to my advantage, as the material was new to most of my peers but was simply a nice refresher to me. My hope was that this would be my key to making new friends and study buddies. I quickly learned, however, that transferring as a fourth-year student and trying to get into well-established study and social groups would not be as easy as I thought.

During class, our Engineering Dynamics professor, Dr. Whiteman, was known to put a practice problem on the board and ask us to work with the student next to us to solve it. I sat on the end of a table in the front row, so there was only one student to my right for me to partner with. I noticed the first time we were tasked with a practice problem my neighbor turned his chair away so his back was facing me and he could work with the guy to his right instead of working with both me and the other guy to his right, as most students in the class did.

I assumed this was just a one-off, so I didn't think much of it. Yet, every class he would repeat this same action, turning his back to me, and I would be left working alone. At one point our professor pointed out that some of us don't work well with others and that is not how real-world engineering works. I was mortified. I just knew he had to have been talking about me. Consequently, I went to his office later that day to explain myself. Before I could finish, he stopped me to let me know he observed what the student next to me was doing and he was not talking about me in his class announcement. His goal was to change the behavior of the student sitting next to me, not admonish me.

The professor also told me I had his permission to move to another seat, just during the practice problem, to work with someone else. Thankfully, two of my Spelman sisters who transferred to Georgia Tech before me were also in this class and sat in the back of the room. Going forward, I would get up during the practice problems to go work with them, since my neighbor did not want to work with me and I didn't know anyone else in the class.

When it came time for our first exam, everyone was on edge. Unlike other classes, this course had only two exams and a final exam. Each exam only had two questions worth 100 points each. As such, getting one question wrong guaranteed a failing grade. My attempts to branch out and find other Georgia Tech students to study with were unsuccessful, so my Spelman sisters and I would meet up in the library to study together.

During one of those study sessions, a gentleman approached us. I didn't recognize him, but he greeted both of my sisters after which point they introduced us. Apparently, he was also a transfer student studying mechanical engineering who had already been at Georgia Tech (GT) for a year. He saw our notes and recognized them immediately, as he had taken the same professor the prior semester. He offered to share some "word" with us. This is when I learned that "word" is code at GT for old homework and exams.

Apparently, shared "word" is a popular, prized possession for study groups, particularly amongst the fraternities and legacy students whose parents also attended GT so they had access to plenty of "word" going back several decades. As one can imagine, it inherently grants an advantage to those who have it, as they have an idea of the type of questions the professor will ask and thus can prepare accordingly.

Thanks to my Spelman sisters' friend, we were finally able to level the playing field. On test day, I felt ready. The questions looked very familiar, so I confidently breezed through that exam. When we finally got our test scores back, the professor informed us that the average score was a 75, and the highest score was a 98. That 98 was mine.

Everyone was asking around amongst themselves trying to figure out who got the 98. To no surprise, no one asked me. After a week of asking each other and realizing it was none of them, a guy asked me after class if I was the one with a 98. I casually nodded in confirmation. His eyes got big as his eyebrows raised to his hairline. It didn't take long for the news to spread across the class.

From that point on, everyone wanted to study with me. Finally, I had other students to connect with. Finally, I felt like I belonged. Knowing that I didn't allow the student who didn't want to work with me to negatively impact my academic performance was just as satisfying as not allowing my high school counselor to adversely impact my performance. Academic stigmas connected to being one of the few women in mechanical engineering, let alone a black woman, did not define me. *I* define me.

13.4   THE PROVE IT AGAIN BIAS

After five long years and five internships in undergrad, I ended up landing a job in the petrochemical industry. I was excited to finally put all the things I learned in school into action. I started as a reliability engineer (RE), which was pretty rare for new hires. Most REs had two to three years of experience within the company in another technical role. Younger Erika would've been intimidated by that fact, but at that point in my life after having done five internships, I had no problem being in a new role and having to pick it up fast to perform well. My love for learning new things and being a high performer was my motivation.

My first real project consisted of looking at maintenance work order data to identify the equipment classes that were having the most failures and thus warranted further scrutiny. Ironically, I did this same analysis during one of my internships in product reliability at an aerospace company, so I was confident in how to do this exercise. When I finally finished the analysis, I created a Pareto chart showing which equipment classes had the most failures, along with a one-page slide giving some background on the methodology I used.

I was excited to present these findings to my business team, the people who asked for this information. However, during the presentation, the more I spoke, the more I noticed furrowed eyebrows, squinted eyes, and looks of confusion across the room. After the presentation, my excitement died a quick death when I was bombarded with questions and doubt surrounding my results. "How would piping have the highest failure rate? That's inconsistent with what we see in the field." "Pareto charts really don't make sense in this application; a pie chart would've been better." "This doesn't look like the right data, where did you get this data from?"

In the end, the business team leader asked that I go back and redo the analysis. I smiled and happily agreed, but inside, my blood was boiling! This was not my first time doing this type of analysis, so why would they have this much doubt? Sure, I was new, but having interned at this company before, as well as plenty of other companies, and having math and engineering degrees, I was confident in this very simple analysis. However, my confidence didn't matter here. What mattered was the fact that I needed to prove my results—again.

I was forewarned by the Women's Interest Network (WIN), one of the company's many employee resource groups, about this gender bias before: the *prove it again bias*. This is a bias faced by women where they are forced to prove themselves repeatedly throughout their career, their success is discounted, and their expertise is questioned. Men in technical fields are granted the benefit of the doubt, while women, especially women of color, begin with the deficit of the doubt.

The business team's perception of my results was not in my control. However, my performance and ability to repeat the analysis was in my control. Not only did I redo the analysis, but I also made a longer slide deck walking through each step in the methodology, as opposed to simply presenting a one-page slide of the steps. I included screenshots of how I extracted the data to help ease any doubt surrounding my data source. I included an explanation of how I cleaned and grouped the data into equipment classes, based on the data source, to ease doubt surrounding which equipment class had the most work orders. I included a slide explaining how to interpret Pareto charts and the value they have in business decision making to help ease any doubt surrounding why I knew a Pareto analysis was best fit for this application.

While looking at a screenshot of the data, the business team leader noticed that several of the work orders were written to the wrong equipment type. This explained why my initial results showing the top 20% of equipment classes with problems were not representative of what they saw in the field. The maintenance planners had been incorrectly assigning work orders, leading to incorrect analysis results. There was nothing wrong with my initial or subsequent analysis; errors in their work process were the root cause.

The value of this analysis quickly spread across other business teams, as they too wanted a Pareto analysis of their unit to identify the top 20% of equipment classes with problems. Consequently, my boss asked me to develop and facilitate a training for the other reliability engineers so they could do the same analysis for their business teams. Knowing that I didn't allow the business team leader who doubted my results to negatively impact my career performance was just as satisfying as the issues faced with the student in undergrad and the counselor in high school. Professional stigmas and doubt connected to being a woman in a male dominated industry, let alone a black woman, did not define me. *I* define me.

13.5 PROFESSIONAL STIGMAS AND DOUBT DO NOT DEFINE ME, *I* DEFINE ME

Four years later, after having rotated through three different engineering roles, I became an equipment strategy facilitator, the individual responsible for building a risk-based plan around how to maintain and monitor the equipment to get reliable performance. At this point, I was more than familiar with petrochemical equipment and how to maintain it, so I knew I could excel in this role.

During a discussion with one of my maintenance leaders surrounding an upcoming equipment strategy inspection, we did not agree on the due date. I watched as he explained why the inspection was due too soon and needed to be pushed out to a later date, based on the risk of asset failure. Once he finished, I tried to explain how I calculated the risk and why the due date aligned with said risk, per company policy. However, before I could finish, he cut me off and told me to "calm down, it's just a discussion".

I was instantly confused because I was not upset at all and yet somehow, that was his perception. My tone was not elevated, my posture had not changed, I was not waving a finger in his face, I was simply being myself and explaining how I reached the resulting due date I selected. Like the former gender bias, I had been forewarned by the Black Employee Success Team (BEST), another company employee resource group, about this racial bias: the angry black woman. This negative stereotype unfairly characterizes black women as being more hostile, aggressive, and contentious, and having an overall angry personality. This same stereotype is not applied to men or other groups of women, only black women.

As the maintenance leader cut me off to try to calm down my non-existent anger, he continued to defend his side of why my date was not right. I allowed him to finish before politely explaining that I was not mad at all, that this was just a normal conversation with no cause for flared emotions. I also tried to help him understand that I have a naturally deep voice that may be different from the women he's used to being around that have a more feminine voice.

While we didn't reach a solution we could agree on, I felt good that I had the courage to speak up for myself. I didn't realize at the time how much the support of WIN and BEST helped prepare me for these type of adverse career headwinds. Knowing that I didn't allow the business team leader who doubted my results nor the maintenance leader who questioned my emotions to negatively impact my career performance was just as satisfying as the issues faced with the student in undergrad and the counselor in high school. Professional stigmas and doubt connected to being a woman in a male-dominated industry, let alone a black woman, did not define me. *I* define me.

13.6 TRANSFORMING HEADWINDS INTO TAILWINDS

In every example recounted here, similar themes appear in my approach to addressing headwinds tied to my social identity. The first key is understanding what is in your control versus what is not, just as my mom reminded me in high school through the serenity prayer. Situations that are beyond our control present themselves as headwinds, barriers to our success.

However, by changing our perspective and accepting what we cannot change, it allows us to channel our focus to that which we can influence and control. Additionally, upon understanding what is in your control, you can now harness that same energy that presented itself as a headwind and use that as your motivation to excel. Stigmas associated your social identity do not define you. *You* define you. You can successfully transform the headwinds into tailwinds by changing your perspective and understanding what you can control.

Secondly, finding connections in these spaces that are not as welcoming helps to combat feelings of isolation and self-doubt that can be common in non-inclusive environments. In high school, having my English teacher invite me into her circle with her daughter changed my feelings of loneliness. In college, having my Spelman sisters and their connection with another experienced student helped not only to change my feelings of loneliness, but also to combat the festering self-doubt that I even belonged in that engineering program.

Having employee resource groups to network and supply career advice helped prepare me to address the doubt of my abilities and my emotions that could've been headwinds in the absence of connection. While it may be difficult to form those connections on your own, finding clubs with similar interests if you're in school, or employee resource groups if you're in a professional setting, can help facilitating those connections.

Lastly, it's important to realize that just as individuals who belong to intersectional groups can change their perspectives to turn headwinds into tailwinds, our allies, people from majority groups, are equally capable of helping to remove headwinds altogether. As Fiona Byarugaba explained, "An ally is someone who is not a member of a community but engages in action to support said community". Active allies are those who witness injustice or unconscious bias and work to address it to drive equity.

This is what my high school teacher did when she realized I didn't have any friends. This is what my college professor did when he attempted to get another student to work with me and gave me permission to go work with someone else. This is what the company employee resource groups did by having seminars to teach employees how to recognize and address bias. In order to step in and promote equity and inclusion, allies must first know what intersectionality is and how to recognize the headwinds that can accompany it. While my story highlights the headwinds imposed primarily by my socioeconomic status, gender, and race, there are many other social identities referenced earlier that are also not granted privilege, in one form or another, like the majority is.

Utilizing the foregoing techniques helps transform headwinds into tailwinds, allowing us to excel in spaces that were never designed for or accepting of people from non-majority groups. Further, building up the army of intersectional allies will ultimately lead to the biases and negative behaviors directed toward intersectional groups to dissolve altogether. Diversity breeds innovation, and STEM is driven by innovation. Consequently, if we as a society aim to continue making pioneering advancements in STEM, we must embrace and make space for different social identities and recognize the value added when everyone gets a seat and a voice at the table.

ABOUT THE AUTHOR

Atlanta native **Erika Anderson**, MS, is a self-proclaimed education enthusiast who graduated with honors with bachelor's degrees in mathematics and mechanical engineering from Spelman College and Georgia Institute of Technology, respectively. Upon completing her undergraduate degrees, Erika began her career in the petrochemical industry. After rotating through various engineering roles, she found reliability engineering to be her niche and has thus been growing her skill set in this area ever since. Six years into her career, Erika heard the call of education once again and happily answered by pursuing a master's degree in analytics at Texas A&M. Shortly after completing her graduate degree, she transitioned to the paper and packaging industry, where she works today as an Asset Health Leader ensuring reliable operations of paper mills and converting plants across the United States. Although engineering feeds her wallet, Erika always emphasizes that STEM outreach feeds her soul, as it is her true passion. Due to her STEM-related community involvement over the years, she was selected by the American Association for the Advancement of Science in partnership with Lyda Hill Philanthropies to be an IF/THEN Ambassador. The mission of this organization is to expose more girls to STEM fields because it is known that if she can see it, then she can be it. As one of 125 women in STEM chosen to be Ambassadors, Erika has been featured on television and in books, and even has a full-sized statue that was featured at the Smithsonian Museum along with the other IF/THEN Ambassadors.

Erika doesn't have any kids, but she is the proud dog mom to Ladii, her Australian shepherd, adopted by her and her late husband, Wick Anderson, 10 years ago. Although her husband is no longer here physically, Erika takes pride in knowing that she is cultivating the seeds they planted together by furthering their mission to prepare scholars for academic success through STEM outreach. When passion meets purpose, the opportunities are endless!

Although Erika is no longer physically here with us, her legacy lives on in the people who she continues to inspire with her story.

14 The Journey to Becoming the First Wellness Engineer

Lennis Perez
Just Lennis LLC, USA

When the doctor throws the word *cancer* as part of your diagnosis, your life gets turned upside down. A memory from the summer of 2014 I'd never forget.

As someone who's very analytical (a trait that most engineers have), I felt an urge to figure out what had happened that led to that diagnosis.

After recovering from the initial shock of the news and going through the recommended treatment, I found myself constantly thinking and asking the same questions over and over again: "How can this be? How did I get here?" This experience forced me to evaluate my entire life.

I'd like to take you back through the years prior to that diagnosis and share the process that led me to where I am today: a healthy, strong, and fulfilled human who has made it her mission to help others find their most authentic selves through well-being.

14.1 THE PATH INTO ENGINEERING

I grew up in Venezuela, an oil-rich country. When I was young, my Dad worked for the army and was a civil engineer, my Mom worked as a kindergarten teacher, and my older brother was pursuing engineering by the time I was in high school.

During my senior year, my parents hired a tutor to help me with math and chemistry. He was a chemical engineer and loved his career. This relationship had a positive impact on my confidence to pursue engineering.

I felt pretty ambitious at that young age and wanted to become a top executive at a big company. When I found out that many CEOs at the large oil companies had an engineering degree, my future became clear!

I told myself, "That's it! I'm going to study chemical engineering and work for the oil and gas industry."

I was accepted into one of the top universities in Venezuela – La Universidad Central de Venezuela (UCV). It was an exciting time, up until my Dad got assigned to work for the United Nations in New York City as a military advisor. That meant my entire family was moving to the United States for the next two years, right as I was finishing my freshman year of chemical engineering.

DOI: 10.1201/9781032679518-16

14.2 LOOKING AT THE BRIGHT SIDE

We moved to the United States in October 1999. I decided to put my degree on hold so I could learn English before returning to Venezuela to finish my chemical engineering degree. But life had other plans.

While living in White Plains, New York, I got accepted into a community college and started taking first-year engineering courses (again). That's when I learned that you could graduate as an engineer in the United States after going to college full-time and finishing a four-year course curriculum. In contrast, to become an engineer in Venezuela, the course curriculum is five years, and you have to complete a thesis.

I did great in school and ultimately transferred to Manhattan College, where I got a partial scholarship to finish my bachelor's degree and a full ride to get my master's in chemical engineering. But as an immigrant, I needed a sponsor to stay legally in the United States. I was able to secure sponsorship, and my professional career as a chemical engineer officially started in 2005.

I worked for a company that designed plants for the oil and gas industry, from refineries to ethylene and polypropylene plants. The first few years were great, but right before my five-year anniversary, I found myself evaluating my career trajectory within that organization.

I could see myself on two clear paths: either becoming a technical expert or going into management and sales. But, I had this inner feeling that I couldn't see myself in the shoes of the leaders in that organization. In my eyes, they seemed to be living "Groundhog Day" as many were stressed, tired, and constantly running around. Others traveled a lot, and I can't think of any managers who actually demonstrated a good work-life balance.

14.3 MAKING THE SWITCH

I recognized that I was good at designing chemical plants and wanted to go deeper into developing my technical skills. So, after five years, I switched companies. At the beginning, I LOVED my new job; I was learning so much, I worked with a great team and I had lots of responsibilities. I felt my colleagues trusted my opinions and work.

On the personal side, I had gotten married and had purchased a house. I was living life as I was supposed to. But right before my 30th birthday, I found myself asking, "Is this all there is to life?" "Am I supposed to do this for the next 30–35 years and then retire and die?"

These questions brought up some red flags to my husband, who recommended that I see a therapist. I felt so ashamed when the therapist shared her diagnosis: I was suffering from anxiety and depression. Through the many sessions we started to uncover the root cause of my anxiety and depression. The therapist explained that all the "negative" emotions I pushed back through some major life events had taken a toll on my mental health. I never grieved when I moved from Venezuela to the United States, or when my parents moved back to Venezuela.

I remember clearly as they were packing up the apartment in White Plains, sitting in my room and just bursting into tears. My parents came in and my Mom gently said: "You don't have to stay if you don't want to" as she tried to hold back her own tears while my Dad just sat there in silence. I vividly remember thinking to myself: "You can do this, this is the best decision for you. It was the most logical decision, that's it." I told my parents: "I'd be alright, I know staying in the US is the best decision for my future." Through therapy I also discovered my inner battle with perfectionism, which I confused with ambition and having high standards for myself.

Mental health issues were not uncommon in my life. Two of my aunts have mental health challenges, which I heard of as I was growing up (even though my family tried to shelter my brother and me from these conversations). One of my aunts suffers from clinical depression, and my other aunt has schizophrenia. Mental health topics are taboo in my culture, but it was through those experiences I had heard when I was young that at least I knew the difference between a psychologist and talk therapy versus seeing a psychiatrist.

I didn't want to jump into medication straight away, I wanted to first explore what was happening inside and then if medication was the next step, I'd have to make the decision then. Since my depression and anxiety were on the mild spectrum, talk therapy seemed to be helping. After a few sessions, I started to feel better, but I still struggle with my perfectionism. I wanted to be a perfect employee, colleague, wife, and daughter, and show up at 110% for everyone else. I had completely overlooked the most important person I needed to show up at 100% in my life: I had neglected myself.

14.4   THE TRIGGER

Fast forward to the holiday season of 2012, and this is where my world truly started to fall apart. My husband and I sold our first home and purchased a house in a beautiful neighborhood in New Jersey. My parents were visiting for the holiday season and helping us remodel our new home, but a few days into their visit, my Dad ended up getting sick and hospitalized for 45 days.

For weeks, my days consisted of going to the hospital before 7 a.m. to check on my Dad's state and talk to the nurses. I was usually translating for my parents, as they don't speak English. Then, I'd head to work around 7:45 a.m., drive to the hospital during my lunch break, head back to work, leave around 5 p.m. to go back home, and work on the house until 11 p.m. or midnight.

Some nights I would stay in the hospital so my Mom could rest at home, other times my brother would drive down from Boston to stay the weekend with him. Even though it was a family effort, I took the role of "captain of the ship", controlling everything and pushing through. It was one of the hardest things I've ever gone through in my life.

By March 2013, my Dad had gotten better, he went through surgery and by May, just a few months after surgery, my parents traveled back to Venezuela. We finished the house, and I thought to myself that it was finally time to enjoy life again. Which brings me to that diagnosis I shared at the beginning of my story during the summer of 2014.

After going through those periods of high stress and getting completely burned out, my body was screaming. It was saying: You can't continue living like this.

14.5   A HUMAN DOING MACHINE

The cancer diagnosis was a true turning point in my life. I felt I had two choices: remove myself from the stressful environment, or continue to push through. The second choice was what I had been doing for over a decade, and it didn't seem to be working anymore.

I recognized I had fallen into the "Groundhog Day" trap myself. I waited anxiously for weekends and vacation days so I could enjoy myself. But during my marriage, I spent most weekends going to races to support my husband as he competed in triathlons, swim races, and other sport activities.

I had completely lost myself through my adult journey. I had been focusing so much on *doing* what everyone expected me to do, on living the "perfect" life by being perfect to everyone, that I didn't know who I was anymore. The person I had become through the years was so foreign to me, I didn't want to be her anymore.

I decided to leave my marriage, as I was learning it wasn't a healthy relationship. There was a lot of codependency and poor communication. We simply stopped listening to each other and our egos were getting the best out of each of us on this "partnership".

Just as luck would have it, the following month after the divorce was finalized, I was laid off. I took this experience as a sign: it was time to turn off the "human doing machine" my life had been for decades.

14.6   FINDING ME – THE HUMAN BEING

Now what?

I had built enough financial stability to give myself the space and time to explore new passions and continue to learn about who I truly was. I invested in coaches, I started a food blog and YouTube channel and attended my first food bloggers conference. I got very curious about what the world looked like outside of engineering.

After six months of exploring this new world, I was hired back to work in a different department at the same company that laid me off. But, that new fire within was still very much alive. I started taking courses on things I was passionate about: nutrition, gut health, positive psychology, mindfulness, and meditation.

From 2016 through 2020, I switched companies, and my new job allowed me to travel to Europe and South America. In that period of time, I fell in love again and also became a caregiver for my father, who now lives with my partner and me full time.

Then COVID happened. An event that opened my eyes to bigger possibilities. It gave me the courage to step into my most authentic nature. I started coaching and helping other women in STEM learn the tools to manage their own stressful lives. Then, opportunities opened up for me to speak at different organizations about the importance of employee well-being. This is how I became a wellness engineer.

14.7   THE WHOLE HUMAN APPROACH

As a wellness engineer, I teach practical and sustainable solutions for those I work with. Through this work, people get to experience firsthand what their lives can look like when well-being is a priority on their to-do list.

Here's the process I teach many of my clients:

The first step is to RECOGNIZE that you are much more than your job title
    and job performance.
Finding gaps in your weekly schedule to detach yourself from your day-to-day
    work that pays the bills is key, even if you work your dream job!
The next step is to EVALUATE where you are in life and spend some time
    reflecting on how that feels. You can grab a piece of paper and answer the
    following questions:

- Am I content and joyful most days?
- Do I dread the weekdays and wait impatiently for weekends and/or
    holidays?
- Am I battling constant energy lows or health issues?
- Do I feel I have a healthy way to cope with stress that doesn't involve
    "numbing" mechanisms? (numbing mechanisms include procrastinat-
    ing, overcommitting, avoidance, addictive patterns, etc.).
- Am I honoring time to fill my own cup?

Answering these questions honestly, even when it feels uncomfortable, allows
you to see the things that may not be working for you right now.

The last step is to DECIDE. Many times we forget we can decide what's next for
us. This can come as getting some courage to ask for help, or taking a career break.
It boils down to prioritizing YOUR happiness and overall well-being, even if it
sounds ridiculous to your inner circle of friends and family.

14.8 WHAT'S NEXT FOR YOU?

Life is more than our intellect and accumulating knowledge. Life is experiencing
the ups and downs of being human. Life is about integrating all the emotions we go
through, even those that are uncomfortable and even painful. Finally, life is about
staying curious and connected to that thing within you that warms your heart and
makes you feel giddy.

This is how you bring wellness into your life – by balancing your intellectual
quotient (IQ), your emotional intelligence (EQ), and your spiritual intelligence (SQ)
such that they all work in equilibrium.

Even though my journey started in a different country from where I live, with a
completely different plan, I'm grateful for all the challenges and roadblocks, for all
the unexpected twists and turns. They gave me the lessons I needed to build the courage to evolve into who I'm meant to be.

Take the following as a gentle reminder:

Living Life Using the Whole Human Approach: physical health, mental health,
    spiritual health. A circuit that needs to operate in balance to bring fulfill-
    ment and peace into your daily life. You're worth it.

ABOUT THE AUTHOR

Lennis Perez is the founder and chief wellness engineer at Just Lennis, LLC, a wellness consulting company focusing on helping organizations and individuals implement stress management strategies to prevent burnout and improve efficiency, productivity, and overall well-being. Through her workshops, events, and individual mentoring, she teaches professionals in STEM how to manage stress in practical and sustainable ways to ultimately become the conscious leaders and role models of well-being in their workplace. She has a Master of Science in chemical engineering with over 16 years of experience in industry. She is also an international public speaker, certified meditation teacher and coach. Combining her master's in chemical engineering with her meditation and coaching practice, she brings a unique but relatable approach to wellness in STEM.

15 Dr. Kenya's Nonlinear Journey

The Intersection of Science, Education, and Policy

Kenya L. Goodson
Hometown Action, USA

Scientists and engineers often approach problems with a linear mindset, seeking straightforward solutions that follow a predictable path. However, we must recognize the value of thinking nonlinearly and embrace the intersections where different ideas, disciplines, and experiences come together. Life rarely follows a straight path and is laden with successes and failures. However, our experiences and decisions intersect unexpectedly, leading us down new paths and shaping our purpose and destiny.

As a result of globalization, jobs are more competitive. Some new graduates work in jobs which are unrelated to their education or training, but those experiences can provide a unique brand and niche in their field. By embracing the intersections between seemingly unrelated experiences, they can create new opportunities and find success in unexpected ways. I worked at places that had no relation to my degrees. But my journey provided extensive experience and expertise, giving me a unique brand—an intersection of environmental science, education, and advocacy.

This chapter explores my career journey. I want readers to understand that our lives shift in many ways that take us by surprise. Those twists and turns are necessary to center us on our career path. Do not be discouraged by unexpected detours. Soft skills are developed by facing and overcoming challenges. All of us experience intersectionality to some degree. Upon reflection, I realize I would not be the person I am today if my life took a standard path.

15.1 MY FORMATIVE YEARS

I grew up in poverty, but I believed a college education would help me to escape my conditions. My family believed this also and emphasized that I needed to go to college. I lived with my grandmother as a young child, with strict guidelines to follow for how I lived. I was raised as an only child despite having four younger siblings, so I spent a lot of time reading and playing with imaginary friends. When I moved with my mother, we were in a blended family. My stepfather and mother lived with three of my younger siblings and three of my step-siblings. It was the most enjoyable part of my childhood. Yet, divorce occurred, and my mother became a single parent of four. Seeing what she endured pushed me to make something of myself.

DOI: 10.1201/9781032679518-17

I had an affinity for math and natural science starting in elementary school. In high school, I liked physical science (chemistry) more than any other natural science. I also performed well in algebra and precalculus. I continued my knack for science as a chemistry major as an undergraduate student at Stillman College, a local HBCU. I do not remember when I learned what a PhD was—maybe in college—but I was determined to receive one because it was the highest degree attainable. Because I was poor, I was ostracized and made fun of. Getting an education was the key to having a better life. Also, I wanted to prove to the naysayers that I was valuable. Life experiences humbled me and taught me a great lesson about true success.

Community service was an essential factor in my college experience. I thoroughly enjoyed my first clean-up, which was a precursor to my career as an environmentalist. However, I was still determining what career path I wanted to follow. I had not met anyone in environmental work, just biologists and chemists. I had yet to discern what my profession should be.

After reflection and encouragement from my father, I decided on an environmental career. It was my first approach to intersectionality because environmental science encompasses different facets of natural science, life science, and social science. I wanted to work in science and with people, hence the environmental profession. It was the most significant decision that I had made.

15.2 THE PIVOT AND DEFINING MOMENT

After Stillman College, I took a 600-level (mistake) environmental engineering class as a non-major at UAB, but I didn't do well in those classes. I believe, in hindsight, that failing that non-major class caused me to fear getting into an engineering program. Why should I be in a graduate program in engineering? I wasn't even an undergraduate engineering major. I wanted to talk to an environmental scientist about the next steps, but I knew no one. So, I moved forward without any mentorship or guidance. Subsequently, I attended Samford University for an 18-month, non-thesis master's program in environmental management. It was a great program to begin an environmental career.

After my master's program, I worked in sanitation for the Alabama Department of Public Health. It was one of the most defining moments of my career. I worked in rural Tuscaloosa County as a public health environmentalist. I inspected onsite sewage disposal and septic systems and issued permits to improve installation. I enjoyed learning about soil profiles and doing site visits. I mostly enjoyed teaching residents how to be compliant with their septic systems. I also saw economic and environmental disparities in support, with many injustices being along racial lines. I wanted to DO SOMETHING. But, I was limited in making fundamental changes, so I shifted my career to do something that made a difference.

I decided to forego my fears, enter an environmental engineering program, and get my PhD in civil engineering at the University of Alabama. I went into a PhD program not realizing that I needed an understanding of how doctoral programs work. I didn't have a PhD coach or even research experience with my master's degree, so I was significantly unprepared. Even though I have cousins who finished their doctoral program, in essence, I was a first-generation PhD of color. Socialization in academia is an essential step to success, and I did not grow up in an environment of academia. Dissertation support groups, my dissertation chair, and my research cohorts helped me to maneuver my process and kept me on track to complete my PhD. This is why I developed a philosophy of helping the underprivileged, who sometimes lack exposure in areas that mostly white men are privy to. I finished my PhD, the first Black woman to graduate with a PhD in civil engineering from the University of Alabama.

15.3 TAKING RISKS AND EMBRACING LEARNED LESSONS

After graduation, I moved to the District of Columbia. I aimed to work for the federal government in research and environmental policy. Getting federal positions is a significant feat, but I worked hard to be added to a cert list for an interview. I didn't get past the gatekeeping AIs used for the applications. I did work for a small minority-owned environmental consulting firm. My first consulting project was researching productivity in stormwater best management practices (BMPs). I gained experience in stormwater quality, but I developed an interest in environmental justice and environmental racism. The EPA had scientists who specialized in environmental justice. I also joined organizations that advocated for environmental justice in the Anacostia watershed. I did not fulfill my goal of working with the EPA. And my career was not going in the direction that I wanted. I decided that it was best to move back to Alabama. It was the hardest decision that I have made in my life.

I failed. I did not find what I was looking for in the District. No government job. No environmental job at all. Failure can be a hard pill when you check all the boxes, cross all the "t's," and dot all the "i's." Failure for me was redirection. Before I left D.C., I started adjunct teaching—another pivotal moment. I loved teaching. I began to discern what I had a gift and a passion for. I LOVED to learn. And I love telling other people what I learned. If that is not teaching, then I am not sure what is.

15.4 MY NONLINEAR JOURNEY

It was difficult, but I moved home to Alabama to start over. Start-overs are alright; just redirection. I gained many things while in the District. I connected with a professional community in the District that supported me and still do. I also had a community in Alabama who supported me, such as my UA colleagues, family, and church.

I began volunteering with various environmental non-profits in the state. Through this, I began understanding some of the environmental ills I remember seeing as a state employee. The structural problems were the same as they were 15 years ago. I started developing my leadership and served as a Board member and later on the Executive Committees for three environmental non-profits specializing in public policy and watershed protection. I served on outreach, environmental education, and EDI committees.

I also returned to my love of teaching. I taught natural science, high school physics, and chemistry as a science instructor and college professor at Stillman College and the University of Montevallo, respectively. I taught in academia for five years with college prep and undergraduate students, exposing them to the world of environmental science using my board connections. I gave them exposure that I did not have as an undergraduate, and I made sure that this included students of color. My goal was to inform them about the biodiversity of the Alabama watershed and some of our environmental issues.

In conjunction with my teaching and outreach, I became civically engaged in 2016. I joined local grassroots political organizations doing voter education and voter registration. I attended council meetings, did volunteer lobbying, canvassing (door-knocking and phone banks), and worked on national and local campaigns. Climate change has centered much of our environmental work because it influences every current environmental structural issue in Alabama and even worldwide. My work in the environmental space is to inform the community of how climate impacts everyday life. I also lobby policymakers about climate policy using my civic engagement experience. I use my education and experience to do what I am doing now.

15.5 CLOSING REFLECTIONS

To address environmental justice and climate change, engineers and scientists have to incorporate a community-based approach. As an engineer who works with community engagement, public policy, and education, I'm proud of all my accomplishments. My career journey is far from being over. I'm proud of the person that I have become. After writing this chapter, I now realize I have a wealth of knowledge and experience. I would not have the perspectives I have now following a traditional path. This path led me to an affirming career and a holistic mindset on how to solve the environmental challenges of today.

ABOUT THE AUTHOR

Dr. Kenya L. Goodson is an environmental engineer, educator, community organizer, and climate policy advocate who focuses on developing solutions to systemic concerns in environmental health, environmental justice, and climate resiliency. She is a native of West End Tuscaloosa, with close proximity to the Black Warrior River. She received her Ph.D. in civil and environmental engineering from the University of Alabama, where she was the first African American woman to earn a Ph.D. in her department. She has acquired over nine years of experience in wastewater management, stormwater quality, and environmental regulation before transitioning in academia and community engagement. She served as faculty teaching environmental science and other STEM courses at the University of Montevallo and Stillman College, respectively. Dr. Goodson is also very involved in the community. She volunteers for several environmental boards. Her volunteer work as a board member focuses on environmental education and outreach, DEI (diversity, equity, and inclusion), and environmental justice. She is also an advocate for climate change legislation and has been lobbying members of Congress in Alabama as a volunteer representative for climate organizations.

Dr. Goodson currently works as Climate and Sustainability Coordinator for Hometown Action/Hometown Organizing Project, where she builds community relationships to organize for climate resilience initiatives in rural communities impacted by climate-related disasters.

Section III

Leading the Way

16 Women Redefining Work-Life Balance in STEM

Integrating Parenthood into the Equation

Sarah Marie Bilger
IPS – Integrated Project Services, USA

You know those jokes that start out saying, "*A priest walks into a bar …*", or maybe something along the lines of, "*Why did the chicken cross the road …*"? These jokes usually pan out because you can visualize them. You're able to take those few words, conceptualize an image, and let the joke take life within your own imagination. Your lived experiences and knowledge of the world allow these jokes to make sense.

So now, I'd like you to take a moment and imagine the joke starting off by saying, "*An engineer walks into the room …*". What does the engineer look like to you? What are they wearing? How do they act? What specific distinctions make them an engineer? Did you imagine a woman? Is she standing no taller than five feet tall? Does she have a ring on her finger? Is she married? Does she have two young kids, one propped up on her hip, the other pulling at her clothes trying to get her attention? No? That's not what you imagined? If you told me that same joke, I wouldn't imagine that person either. Would you be surprised to know that woman is an engineer? Would you believe that woman is me?

16.1 THE GENDER IMBALANCE IN STEM FIELDS

Traditionally, most STEM fields have been male-dominated, with women historically being underrepresented in these fields. Although there has been progress in recent years, the gender imbalance we see in STEM fields remains a significant issue, thus creating a gender imbalance among those pursuing and actively working within the science, technology, engineering, and mathematics professions.

According to data from the National Science Foundation, women make up only 28% of the science and engineering workforce in the United States, and the representation of women varies significantly by field of study. For example: women have a higher representation in fields such as scientific research and development, but are drastically underrepresented in fields like computer science, physics, and engineering.

DOI: 10.1201/9781032679518-19

It leads us to wonder why there is such an imbalance within different fields. Is it necessary to see a more balanced workforce, and how would we level out the divide?

Several factors contribute to gender imbalance, including societal biases and stereotypes that discourage girls and women from pursuing STEM careers altogether. Structural barriers like the lack of access to mentorship and networking opportunities, as well as workplace cultures that may be unwelcoming for many women, have been seen to greatly impact the number of women working amongst men in STEM positions.

Presently, there have been initiatives to encourage girls and young women to pursue STEM education and careers. There have also been efforts to increase diversity and inclusion in STEM workplaces through measures like diversity training, mentorship programs, and policy changes to support work-life balance and greater flexibility. However, we are still seeing low numbers in the industry. As women continue in their fields, the greater issue becomes: how do we continue to encourage women to remain in STEM fields? The disconnect between uplifting our youth and remaining committed towards the mission of encouraging women in STEM needs to extend past childhood activities and college involvement. We need to ensure it's a lasting impact that follows the woman through her career.

16.2   CHALLENGES WOMEN FACE IN STEM

I clearly cannot speak for all women in STEM, but my story is not unique to those on the journey to become an engineer. In grade school, I had always been exceptionally gifted at math, typically taking the most advanced classes in my grade, and doing well year after year. Science class always had a way of keeping me intrigued, and I enjoyed all the labs, experiments, and discoveries that were covered all the way up through high school. When we began taking college aptitude tests, it seemed like the next logical step that I would pick something related to math and science.

At the time, my grandfather had a large influence on me as well. He would cut newspaper articles and scribble down notes just for me on different career paths to choose from. After his input, reading a few books from the school library, and talking it over with some teachers, I had decided to go to college to major in engineering.

I had never met an engineer before, or even really knew what the major would entail. I never recalled many shows or movies which specifically called out an engineer in it. I vividly remember my younger cousin actually asking me, when I became an engineer, if she could take a ride on my train. Clearly, she wasn't sure either. I knew it wasn't going to be that type of engineering, but my perception of what a day-to-day life could possibly look like for an engineer, or all the different disciplines, was non-existent. I certainly was not aware of the lack of women that were in the field, until I started college.

Uniquely for me, I had an unusual upbringing that others necessarily did not have. Growing up, I was mainly raised by dad. My mom passed away from cancer when I was 9 years old, and the influence of my dad allowed me to develop a kind of no-BS attitude when it came to work ethic and independence. If I wanted to go somewhere, I needed to know the directions to get there, including street names and landmarks. If we were out somewhere, I ordered my own food. And I talked to who was in charge to partake in activities like bowling, arcade games, and other purchases. I filled out my own paper-work for school every year and eventually did my own back-to-school shopping as well. While these skills may be developed through a single mother's upbringing, or two parents raising a family together, I do think my experience of losing my mother at such a young age altered my perspective and changed the trajectory of my life's path.

The week before I officially started college at Penn State University, I found myself placed in an organization, and surrounded by other girls majoring in engineering. All of us were anxiously awaiting our first class, bonding over icebreakers and activities to motivate us and give us a sense of community within the group. I left that week feeling empowered and excited about the years to come, only to head out to one of the satellite campuses to be in a group of only a handful of girls in engineering. My very first engineering related class had only three girls in it, and the odds didn't increase after that. The good news was, we knew each other; but the bad news was, we were clearly underrepresented.

To say the least, after many exams and assignments, sleepless nights, challenging moments, encouraging times, joys, and disappointments, I eventually graduated from the main campus, and officially had a bachelor's degree in energy engineering. In my super-senior year, I met my now-husband, who was also an engineering major. He graduated with a bachelor's in mechanical engineering, a semester before I did. We started our first jobs, moved in together, got married, moved from Pennsylvania to South Carolina, started a family, purchased our first home, continued to grow our family, and now here I am, still in the field, still working full time, and being a mom of two, learning how to navigate it all.

The impact of these challenges on women's careers and the industry as a whole contributes far greater into our community. We can highlight experiences of women from diverse backgrounds, including women of color, LGBTQ+ women, women with disabilities, and women from low-income backgrounds. It's important to acknowledge that not all women face the same challenges and barriers in STEM, but that understanding the need to address their challenges, and collectively listening to each other's accomplishments and struggles, can reach far beyond just the women in STEM.

16.3   CREATING INCLUSIVE WORK ENVIRONMENTS

The role of mentorship and networking in empowering women in STEM could shape those entering the field in a positive direction. Similarly to how I started college with an organization built to create mentor-mentee relationships and strategies for building and sustaining a strong network in STEM, these opportunities need to continue as we enter the workforce. However, this does not need to be solely in the representation of women alone. We can learn from all disciplines, all backgrounds, and from all individuals. Creating an inclusive work environment starts with including everyone.

The impact of unconscious bias on women in STEM undoubtedly exists. Think of the joke scenario called out earlier, where we imagine an engineer. Our first thought typically does not go to an individual who would identify as a female. Developing awareness and strategies for addressing this can start to shift the paradigm of what it truly means to be an engineer or, even more broadly speaking, an individual in a STEM career. Training programs for managers and colleagues, as well as creating a culture of awareness and accountability, are already being implemented to start adjusting unconscious bias, but we certainly have room for improvement.

I personally have been mistaken for a secretary or admin in meetings from individuals that did not know me. When out on a site visit, I was thought to just be there to take photos and notes. It was not understood that I was also the engineer on site. It sometimes takes individuals a second to process that I am an engineer. While this mistake is common for women in this field, I do not feel that it is a mistake typically assumed for men in the same role. I know I am not alone in these situations. I have heard stories from other women of similar experiences and have heard rude and derogatory comments made towards women on construction sites and in the office, even after trainings and zero-tolerance regulations have been placed.

We can also create inclusive work environments by demonstrating support for women entrepreneurs in STEM and making resources available to support them in their pursuit. This could include information on funding sources, creating business incubators, and holding networking groups specifically for women entrepreneurs. In order to increase the number of women in leadership roles, which would allow representation to be seen within the STEM field, support needs to be provided by all individuals working in the field and far beyond it as well.

In continuation, highlighting successful women in STEM and sharing stories of women who have achieved success in STEM fields, including how they overcame challenges along the way, is essential in paving a path moving forward. This could include interviews or profiles of prominent women in STEM, and their advice for aspiring women in the field. It is becoming more common to see this level of awareness being spread. Resources such as this book are able to create containers for stories to be shared that can exist for generations to come.

16.4   INTEGRATING WORK AND FAMILY LIFE IN A STEM CAREER

In an industry that is still mostly male-dominated, it is vital that we create flexible work environments to accommodate family growth to retain women in STEM. The desire to raise a family is not unique to STEM fields, but because of the gender imbalance typically found, the needs of a woman going through pregnancy or raising children have a greater potential to go unmet. When these issues arise, sometimes the mother may choose to leave her career completely. Aside from having acceptable family leave so that they can properly feel prepared before returning to work, more accommodations need to be addressed. Allowing open conversations to improve overall mental health, providing safe and acceptable spaces for mothers to pump and rest while at work, and implementing flexible work schedules to accommodate family needs are some ways we can begin working together to improve the future and empower women to stay in STEM.

A career in STEM can be challenging enough. But, facing stereotypes, assumptions, and being put in boxes by others or even by ourselves, women are more likely to experience challenges when pursuing a career in STEM. When we add on parenting, we can easily add overwhelming and stress to the equation. We should be able to actively engage in our family lifestyles, as well as advance in our careers, to successfully integrate work and family life.

We push for girls to engage in STEM, and we encourage them to strive for positions in the workplace, but we sometimes forget that it can be done alongside raising a family. Encouragement needs to be placed there as well. We want generations after us to see that it is possible to have these opportunities available. By providing this inclusion in the workforce now, we can inspire the future to be drastically different. We can demonstrate persistence and push for more capacity in the industry to sustain fuller lifestyles.

16.4.1 THE IMPORTANCE OF A LACTATION SPACE

When I became pregnant with my daughter in 2018, I started a new position in mechanical engineering in South Carolina. Thankfully, a co-worker was also pregnant and a few months ahead of me, and I followed her lead. We chatted about feeling exhausted at the end of the day, and experiencing morning sickness along the way. We discovered what maternity leave looked like, and we even navigated the postpartum time period together, after I returned from a 9-week maternity leave.

Before her return to work, there was no designated lactation space in our whole company. Not only was it lacking in the engineering department, but it was not set up for others in our company as well. She had requested a space with no glass doors or windows others could see in, a mini-fridge to store our milk, somewhere to sit, outlets to access for pumps, a sink, and a few extra basics. Slowly, a space came together, and more items began to be added. The sink was out in another room close by, so a few adjustments needed to be accounted for; but before you knew it, four of us were in there using the space throughout the day. We would try to rotate and be aware of each other, but most of the time there were at least two of us in there, since we all worked roughly the same schedule. Although it was not perfect, the ability to have women in the same season of life to talk to every day was a priceless bonus to my time at work; but knowing what I know now, I believe it was rare to experience.

The need for open conversations about parenting in the workplace is vital, not only for the mental health of parents, but for general well-being. Providing these safe and acceptable spaces for mothers to pump and rest at work should be mandatory, but as of now it is not. I love seeing a lactation room on a floor plan when starting a new project at work; however, it's not required. A company needs to place these rooms in their building plans, and to designate them as such.

Aside from opening up conversations about parenting in the workplace, and having lactation rooms, allowing flexible work schedules to accommodate family needs is another great way to create a more family-friendly work environment for parents. I've been seeing more places that offer flexible work-from-home policies, and I have even seen adjustments in personal and sick time off. These accommodations are certainly helpful, but they are still only the tip of the iceberg when it comes to the reality of being able to fully integrate work life and family life into an everyday practice.

16.4.2 FINDING MY OWN UNIQUE OUTLET AFTER BECOMING A PARENT

I returned to work exactly 9 weeks after my first daughter was born. I was given exactly 8 weeks due to an unplanned cesarean section and took one more week of my own paid vacation time. The scar from my cesarean was still freshly healing, the routine of breastfeeding was still feeling new, and the raw emotions of learning to leave my baby for the first time were sinking in. That first day back, I sat at my desk trying to simply remember my login password amongst the many other thoughts that were playing out in my head.

Even with supportive co-workers, pumping accommodations, and having people to talk to, introducing a family lifestyle alongside a career takes on a new layer of complexity. Raising a family and being a mother is the most challenging and rewarding job I will ever have, but my engineering job is important as well. It also requires dedication and a certain level of mental capacity. Unfortunately, we cannot be at two places at once, but we can learn to adjust our perspective so we can be present in both. To make this possible, I needed to seek resources outside of work to assist in navigating life as a working mother.

I began creating the spaces that I craved. I started regularly going to therapy for the first time in my life; I needed to find myself outside of being an engineer and being a mother. I needed to find my own unique outlet to creatively explore. I went back to painting like I did when I was younger, I started running again, I colored, and had favorite TV shows that were outside of children's cartoons. I was able to reboot and recharge outside of work so that I could arrive at work ready to take on the day, and leave feeling accomplished enough to have energy to provide back to my family and the extra activities and hobbies that filled my cup, if you will.

Unfortunately, I was placed on furlough from that position for a few months at the start of COVID. After crying and comparing myself against others, I found myself in an opportunity that was more aligned with my vision of the work-life balance I craved. I was able to work completely remotely, and was back doing the work I knew how to do and felt confident in doing. What felt like a huge disappointment ended up being a wonderful opportunity that has provided so much freedom in my ability to work and raise a family. I learned to adjust and pivot forward, and it reaffirmed that other possibilities were available for me. I had opportunities because of previous experiences and my engineering degree.

Since starting at the position I currently hold, I went on to have another child. Our second was born by an unmedicated VBAC (vaginal birth after cesarean) in January 2022. This time around, we adjusted our finances so that I would be able to take a full 12 weeks off for maternity leave. On top of this, and having the ability to work fully from home, our sweet little boy was able to stay home and work right by my side for six whole months. Being able to nurse him in my home office and hear his laughs and coos while I worked made all the difference in returning to work when the time came. Seeing him learn to grow and develop over time more closely made the decision to eventually start him in childcare easier because it felt more like a choice, not an ultimatum, and the transition was so much smoother. The support and understanding from my employers and co-workers again made all the difference in the transition of welcoming our second child into this world.

I often get asked, "How do you do it all?" Mind you, my husband, who works in the exact same field as I do, does not get asked this question. I honestly think there are those that assume he works, and I stay at home with the children. While it is a perfectly acceptable lifestyle to have one parent stay at home, at this time, it is not the lifestyle we choose. I try to relay to those who ask that this is just a season. The children are young, this does not last forever, so embrace the moment. If you want to be an engineer, and you have both parents who are working full time, a certain lifestyle needs to be adapted.

16.5   STRATEGIES FOR INTEGRATING WORK AND FAMILY LIFE IN A STEM CAREER

A few strategies I've found for integrating work and family life in a STEM career are as follows:

Prioritize: Identify your top priorities, both at work and in your personal life, and make sure you're spending your time and energy on the things that matter most. This can help you avoid getting bogged down in tasks that aren't important, and lead to remaining focused on the things that will have the biggest impact.

Set boundaries: Set clear boundaries between work and home life, and stick to them as much as possible. This might mean not checking work email or taking calls during family time, or scheduling regular time off to recharge and focus on personal activities.

Practice self-care: Take care of your physical and mental health by getting enough sleep, exercise, and healthy food, and taking breaks when you need them. Practice mindfulness or meditation to help manage stress and stay focused.

Utilize flexible work arrangements: If your employer offers flexible work arrangements, such as work-from-home opportunities or flexible hours, take advantage of them to help balance work and personal responsibilities.

Outsource: Delegate tasks that can be outsourced or automated, such as housekeeping or meal preparation. You can also consider hiring a babysitter or nanny to help with childcare duties, if feasible, or utilize full-time childcare, so that you can fully be present with your work.

Communicate with your team: Be open and transparent with your team and co-workers about your priorities and schedule. This can help manage expectations and avoid misunderstandings. If you have a doctor's appointment or an after-school activity once a week, letting your team know ahead of time will help ensure you're able to attend those commitments.

Set realistic expectations: Recognize that achieving integrating work and family is an ongoing process and it may not be possible to always achieve a perfect balance. Setting realistic expectations for yourself and being kind to yourself if things don't go as planned can go a long way in achieving a healthy balance.

While there are ways to navigate the current challenges of developing a successful work-life balance, this creates an opportunity for improvement moving forward for women in STEM. Whether we are scientists, mathematicians, or engineers, or taking on technological advancements, the problem-solving skill sets we possess can help shape a brighter future moving forward for those swiftly coming into the field.

16.6   EMBRACING FAILURE IS PART OF THE JOURNEY

Unfortunately, we don't see many women who are in this stage of life of working full time and having young children. When I was pregnant with our first, I was faced with hearing others ask if I was going to stay home. Now having two, I still contemplate the idea from time to time when projects build up and emotions run rampant at home, but ultimately, I love being an engineer and the purpose I find in having a career.

Engineers often have a reputation for getting caught up with long hours and demanding deadlines. It's not unusual to hear co-workers putting in 50 or 60 hours a week, working weekends, traveling for a few days, and even when the workday stops, they continue to talk about work. This is similar to how I was previously before having children. Occasionally you feel torn between your dedication to your career and the family you are growing. When you're at work, you're thinking about all the household tasks you could be doing, and when you're with your family, you're thinking about the list of tasks you need to finish at work. It can be a vicious cycle to break and is unfortunately common in corporate environments.

In the engineering career trajectory, the common protocol is, after 5 years in a field of study, you can sit for the professional engineering exam. I had reached this 5-year mark soon after my daughter was born, but opted out of taking it immediately since I was still breastfeeding; there are currently no additional accommodations for lactating individuals during an exam. As months went by, being placed on furlough, and getting pregnant with our second, the timeline stretched farther and farther away for when I would take the exam. After finishing up almost a year of breastfeeding my son, I completed the registration process and scheduled out a day to take it.

I ended up failing the PE exam the first time I took it, and I'd be lying if I said I wasn't disappointed. But here I am getting back up, not giving up, and currently still pursuing it all. My story isn't over, and I'm not quite at the point of looking back on all the accomplishments I've made. I can't sit here and officially say "It was all worth it" yet, or give you a story of completion because I'm still navigating it all, still exploring it, and actively sharing it along the way. I'm learning to use my experiences as learning opportunities and to shed light on the paths that are less traveled. If failures didn't happen, and if there weren't bumps along the road, the path would be more heavily traveled, it wouldn't be as special, and it would feel so rewarding.

16.7   INSPIRING THE FUTURE OF WOMEN IN STEM

As current women in STEM, our role in inspiring future generations is to express our individuality and find more ways to open the conversations surrounding women in STEM. Creating more opportunities to gain confidence and cultivate empowerment in girls and young women has the ability to shift the future. Demonstrating that you can accomplish the goals you set out to achieve can showcase resilience and inspire others to see it is possible for them as well.

It's a collective journey, and we cannot get there alone. We can begin by encouraging young girls to seek mentorship, chase curiosity in a variety of interests, and own their power of being able to pivot and learn from their mistakes. If we embrace failure and develop the resilience to get back in and try again, we can change the outcome for what it means to be a woman in STEM.

Looking towards the future of women in STEM, we can understand the impact of inclusive work environments on future generations. There are many ways we are learning to develop a more diverse workforce and highlight achievements made by firsts in their field. We can see the change that is rippling outward from these advancements.

The gender imbalance in STEM fields continues to be an obvious statement. However, the challenges women currently face in these areas of study can be acknowledged and improved upon despite the imbalance. Continuously creating inclusive work environments and pushing the boundaries on how we can cultivate a more integrated work and life environment in a STEM career can inspire the future of women in STEM.

If you are in a STEM field, ask those around you how you can create a more inclusive work environment for them. Creativity cannot flow from overworked and burned-out individuals. A need for flexibility, freedom, and choice must continuously be available. There may come times when you think that this role isn't built for you, and if you continue to perpetuate this imaginary belief that you're not good enough, you will start believing it. You might doubt yourself at times along the way when learning a new program or gaining more responsibility within your role. You may even feel like you want to give up and wonder how in the world are people doing this on a day-to-day basis. You will think that you need to leave the field and seek a different alternative.

But I challenge you to push for change. I challenge you to rethink what's possible in the STEM fields. I encourage you to use your problem-solving skills and your communication abilities to navigate this next stage in your life. You started this journey for a reason; now fight for it, fight to stay. Fight to pave new paths so that future generations can follow your example. My story is still being written; chapters of my personal life journey are still being uncovered; but I know I will always be an engineer. I know better days are ahead for women in STEM. The future is female!

ABOUT THE AUTHOR

Sarah Marie Bilger has eight-plus years' experience in mechanical engineering and holds a Bachelor of Science in energy engineering from Penn State University. She received her Engineer in Training (EIT) certification in 2015 and is working towards obtaining her PE in mechanical HVAC engineering. Her professional experience has taught her several key skills, including managing multiple complex projects simultaneously, working as part of a team, and developing excellent time management and organizational skills. Her multi-passionate personality has led her to become involved with the creation, editing, producing, hosting, and guest appearing for several podcast shows across various platforms including Apple Podcasts and Spotify.

She effectively communicates about her experience and actively sparks curiosity on various topics pertaining to her personal and professional life on her own podcast called *Entering Motherhood*.

Being a woman in STEM (science, technology, engineering, and math) herself, Sarah is highly passionate about others pursuing a career in STEM and advocates for inclusion and representation of women in the workplace. As a new mother of two, she also recognizes the importance of working mothers and the impact it has on our future to be able to accommodate the lifestyles of working parents.

17 Female AND an Engineer?

Exploring the Gender-Occupation Identity Dilemma in Engineering

Helen Sara Johnson
Manufacturing Technology Centre and Loughborough University, United Kingdom

Being female and an engineer can prove challenging, but what role does authenticity play in an engineering career? This is a question I asked during my PhD research. Let me share with you what I found.

This chapter contains insights and experiences from over 300 engineers of what it is like to work in the UK engineering industry. Insights and experiences that will enable you to help yourself and others who work in engineering. This chapter isn't just for female engineers – it's for everyone! And, for those who like to skip to the end to see what happens, here's a sneak preview – yes, you can be female and an engineer. You may need to navigate a few things, but being authentically you is the key to the answer. I want every reader to take something away from this chapter, so I'll also provide some tools to help you on your journey to become authentically you.

17.1 FITTING IN WITHOUT FITTING IN

Fairness was important to me as I was growing up. As a child, especially with a younger sibling in tow, I could often be heard muttering three simple words: 'It's not fair!' Fast forward over 40 years, and, having worked in engineering and manufacturing for over 20 years, the feeling of fairness remains with me; I position equity as central to everything I do.

The male-dominated UK engineering and manufacturing work environments have always felt easier for me. Yet, at the same time, in the past, there were always things that I thought I couldn't influence or resolve – the gender ratio, gendered roles, tasks and behaviours, and the gender pay gap. For as long as I can remember, though, I have battled for identity not to be a predictor of workplace opportunities or outcomes.

Throughout my career, I have also been inquisitive. As my career has developed and I have grown older, I have become fascinated with 'how to be me'. My curiosity led to many questions. How can I be authentic when my identities do not seem to go together? How can I be authentic when I am working in an environment that was not designed for me? How can I be authentic when I am in the minority group? *How can I fit in without fitting in?*

I started talking about my thoughts, feelings, and experiences. And guess what? Other women thought, felt, and experienced the same. I wasn't alone. I started to build a picture of these experiences, each new insight leading to more questions, helping to shape my PhD research further.

I never expected to be offered the opportunity to undertake a PhD, especially aged 40-something. So, when the opportunity came along, I grabbed it with both hands. My PhD has happened because of the fantastic female engineers I have worked with and met over the years. The women who have stood beside me and shouted from the rooftops:

> As females in engineering, we are not the sole note-takers in meetings or tea makers in the office. We are not the social secretary. We require a toilet within walking distance of wherever we work, just like our male counterparts. Most of us are not as tall as the men we work with, and we have feminine curves, so Personal Protective Equipment sized for men will not fit us, nor will it protect us. We do not want to be viewed as your daughter or 'one of the lads'. We do not want to be a diversity hire to fill your quota. And we are tired of consistently working harder than the majority to be noticed and have to fight for our right to be here. Why is it we can be invisible as engineers but hyper-visible as women?

If you are reading this and nodding, remember you are not alone. If you are reading this and having a lightbulb moment (or possibly even doubting my words) – I encourage you to read this chapter to the end. This research comes from a place of good intent as we strive for equity; this research was undertaken to educate others, and this research was conducted to get us talking.

This chapter provides a non-academic insight into my PhD research findings thus far. Throughout this chapter, I want to take you on a journey; to introduce you to the fundamental concepts used in my PhD research – debunking them on the way so we understand their importance within the working environment, not just the academic context that the concepts have been derived from. We will talk about identity, authenticity, and inauthenticity, what they are and why they are meaningful. We'll explore how workplace commitment can be impacted, how careers can be compromised, and how, ultimately, good people leave engineering as they struggle to navigate their identity and authenticity in a male-dominated engineering landscape.

17.2   WHY ARE WE STILL TALKING ABOUT GENDER?

'Why is this important?' I hear you ask, 'Why are we still talking about gender in this day and age?' Let's take a step back – I'll let the statistics tell the story. In the UK, female representation in engineering increased from 10.5 per cent in 2010 to 16.5 per cent in 2022,[1] a 6 per cent increase over 12 years. By my calculations, if every 12 years, the UK increased gender diversity in engineering by 6%, an equal gender split would not be achieved until roughly 2090.

In addition, when you factor in that 57 per cent of female engineers in the UK will leave the engineering profession by age 45, as opposed to 17 per cent of male engineers,[2] gender diversity in UK engineering may not happen in our lifetimes, let alone in this century. Diversity brings competitive advantage, and women play an essential role in our economy as 75 per cent of the world's unpaid work, which subsidises the global economy, is undertaken by women; the total value of this unpaid work is estimated to be at least \$10.8 trillion a year,[3] and when more women work, economies grow.[4]

Circling back to the research, if my curiosity, experiences, and thinking prove right, authenticity and identity interference (the dissonance we can experience between two or more identities) are at the heart of tackling the gender-occupation identity dilemma in engineering. If engineering culture allowed women to fit into a male-dominated work environment without fitting in, we could improve diversity further. Let me explain.

17.3   AUTHENTICITY

Authenticity, being your real or true self, has become a buzzword in life. Feeling authentic at work is a predictor of employee satisfaction and retention.[5] Yet, when 'who we are' does not align with how we perceive our workplace 'wanting us to be', being inauthentic at work may prove more straightforward. When an employee conforms to organisational norms and values that do not align with those of their true, authentic self, a mask or act termed a 'façade of conformity' can be created.[6] This mask is a coping strategy deployed by an employee that enables them to appear that their values align with those of the organisation and, as such, support organisational values. Creating and wearing this mask is exhausting, with impacts on well-being as well as work outcomes. Within the context of engineering, what is fascinating is that creating a façade of conformity relates directly to minority status.[7] In other words, a female engineer is more likely to develop a façade of conformity and suffer the impact of wearing that mask compared to their male counterpart.

Another way to consider authenticity is how our values, preferences, and needs relate to authentic and inauthentic workplace behaviour.[8] When an employee's authenticity increases, so does their job satisfaction, work engagement, and performance. This increased authenticity is also related to high levels of autonomy, learning and development, supervisory support, and colleague support,[9] whilst inauthenticity has the opposite effect.

Authenticity was defined in these two ways within my research – creating a mask and perceived authenticity. Employees may perceive themselves as authentic when workplace behaviours, values, preferences, and needs align. However, when employees conform to organisational norms and values that do not align with those of their authentic selves, a façade of conformity is created, and inauthenticity is experienced. What is less well understood in the work context is what drives perceptions of authenticity at work – this is where my thinking about conflicting identities plays a role.

17.4   IDENTITY

Close your eyes for a moment. Think about your various identities and the many hats you wear. You may be a mother or a father, self-employed or unemployed, unqualified or a university graduate, someone's daughter or brother, someone's carer or lover. However we identify, to feel authentic, we need to perceive that our various selves do not interfere with each other. This can be particularly difficult for women in engineering.

Being in the minority, women stand out. It can feel intimidating when you do not align with the social and behavioural norms of the majority. Feeling like you do not fit in can lead to a lack of confidence, decreased performance, and, ultimately, leaving the industry. Identity interference, the dissonance or conflict we can feel between our identities, can occur when we perceive that our values or beliefs are inconsistent with the norms or expectations of the contextual identity, such as our professional or organisational identity.

This can lead to feelings of tension, dissonance, and emotional distress that can negatively affect individuals' psychological well-being, work performance, and career outcomes.[10] It is plausible that authenticity drives various work outcomes. Still, it is also possible that identity interference plays an indirect effect through authenticity and a direct effect on work outcomes.

Throughout my research, engineers shared their workplace experiences and stories with me. At times these interviews felt like 'Groundhog Day', with the same incidents and anecdotes played out on repeat. I want to share a few of these with you as we explore three concepts or work outcomes which complete our research jigsaw.

Female engineers told me about regular microaggressions and sex discrimination in the workplace and how the rules of engagement differ between them and their male counterparts. They shared about how they have a different offering in the workplace as a woman and how they encouraged and supported other women, yet how some felt the need to use their gender to their advantage and how all these things, even the positives, limit their careers. They explained how they needed to be exceptional to stand out as a minority, mostly thanking Lady Luck for being on their side, but all sharing stories of how they must prove themselves, work harder, fight stronger, and always go the extra mile.

Female engineers described regular features in working life as being talked over, having their ideas ignored only to be regurgitated later by others (and accepted), being undermined, and being told they're 'too emotional' when they display the same passion as their male counterparts. In addition to these experiences were the expectations female engineers perceive others have for them; a female enters engineering, gets bored, has children, and leaves engineering.

It felt like, despite being passionate about the work they undertake, despite their ability to make a difference, and despite wanting a successful career, many female engineers felt they were fighting a losing battle. Many of these female engineers added that they would have told a very different tale if I had interviewed them 3–5 years ago. It was a tale of an organisation they had left, as their experiences were far worse than the recent stories they told me.

Being both female and an engineer doesn't seem to be easy. These two dissonant identities are likely to harm an employee's commitment to work, compromise careers, and ultimately lead to engineers leaving their organisation or, worse still, the industry. As I'll explain, these three work outcomes are fundamental to my research.

17.5   WORK OUTCOMES

We know that commitment is essential for organisational stability and reputation. Associated with reduced turnover costs, more committed employees go the extra mile to support an organisation in achieving their objectives. To measure this work outcome, my research uses affective commitment,[11] the emotion-based 'want' to stay characterised by loyalty, pride, and belonging, as affective commitment generates a higher level of employee performance and reduces employee turnover costs due to increased retention – key performance indicators within the engineering industry.

When an employee adjusts external and/or internal factors, this can lead to the perception that their ideal career goal cannot be achieved without sacrificing other important aspects of their lives. This series of complex trade-offs, known as career compromise,[12] can result in the perception of having a less meaningful career than expected, having less responsibility within roles, making less of a difference or impact, or compromising either role status or career interests.

The ultimate impact of identity interference and inauthenticity is increased employee turnover intentions, the most extreme strategy being 'exit'.[13] When an organisation is not meeting employee needs, or the employee believes their prospects for advancement or success are limited, the exit strategy may be used to cope with workplace dissatisfaction. As a last resort, this strategy includes transferring to a different department or location, resigning, or retiring early.

17.6   THE RELATIONSHIP BETWEEN IDENTITY, AUTHENTICITY AND WORK OUTCOMES

At this point in an academic paper, you typically encounter hypotheses, research methods, and analysis. However, my aim within this chapter is not to baffle you with academic jargon and statistics, but to explain the relationship between authenticity, identity, and these workplace outcomes simply and clearly. Hopefully, the diagram shown in Figure 17.1 helps to achieve just that.

Not only am I interested in the direct paths (arrow b in the diagram) between identity interference and the impact on the engineer, but I am also interested in the indirect paths (arrows a and c in the diagram) from identity interference through the two types of authenticity, as mentioned earlier in this chapter, to the work outcomes.

For context, in the summer of 2022, I undertook a nationwide study across the UK – surveying almost 300 engineers at three different time points over three months and interviewing nearly 40 engineers who had taken the survey. In early 2023, after many hours of work, pots of tea, and the occasional frustrated, teary moment (I had neither undertaken any correlation or regression analysis nor used a statistical software package since circa 1998), my data revealed some fascinating relationships; relationships I would love to share with you.

17.7   MY RESEARCH FINDINGS

Although my PhD is far from completed at the time of this writing, I want to share some highlights from the survey outcomes with you. I want to focus on the non-academic explanations, clarifying what these may look or feel like for a female engineer.

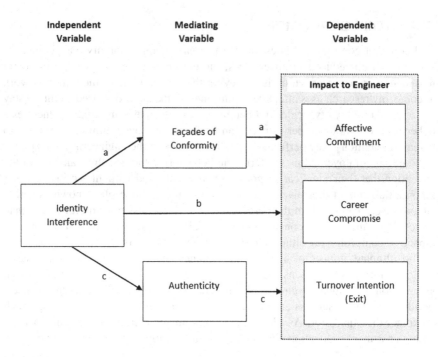

FIGURE 17.1 The theoretical relationship between identity, authenticity, and workplace outcomes.

But first, a short paragraph for the academics amongst us: For the direct paths, identity interference has a negative effect on affective commitment and a positive effect on both career compromise and turnover intentions, with identity interference and career compromise proving the most consistent across the three different time points. For the indirect paths, façades of conformity worked as a mediator in all three paths between identity interference and work outcomes. In contrast, authenticity worked as a mediator for identity interference on two work outcomes: affective commitment and turnover intentions.

17.8   DIRECT PATHS – FROM IDENTITY INTERFERENCE TO WORK OUTCOMES

Back to the real-life, non-academic explanations: we'll take these paths and outcomes individually, exploring and relating them to the engineering world we work in. Firstly, let's explore the direct paths (arrow b in the diagram) from identity interference to the impact on the engineer.

17.8.1   As Identity Interference Increases, Affective Commitment Decreases

Another way of saying this is: as engineers experience greater dissonance or interference between their gender and occupation identity (for example, being female and an engineer), they perceive they are less committed to the industry and/or organisation they work for in terms of their desire or want to stay. For example, as a female engineer, if you experience conflict between what it means for you to be a female and what it means for you to be an engineer, then you are likely to perceive having less emotion or desire around wanting to stay with your organisation or possibly within engineering. This may mean that you feel you do not belong, that you don't feel part of the family, that you may not feel emotionally attached to the industry and/or organisation, or that the industry and/or organisation has lost the personal meaning it once had.

17.8.2   As Identity Interference Increases, Career Compromise Also Increases

Another way of saying this is: as engineers experience greater dissonance or interference between their gender and occupation identity, they also experience more compromise within their career. As an example, as a female engineer, if you experience conflict between what it means for you to be a female and what it means for you to be an engineer, then you are likely to perceive experiencing more compromise in terms of your career; possibly perceiving that you have a less meaningful career than you expected, that you have less responsibility within roles than you want, that you think you're making less of a difference or impact than desired, or that maybe you're compromising either your role status or your career interests.

17.8.3   As Identity Interference Increases, Employee Turnover Intentions Increase

Another way of saying this is: as engineers experience greater dissonance or interference between their gender and occupation identity, their intentions to leave their job, their organisation, or even the engineering industry increase. For example, as a female engineer, if you experience conflict between what it means for you to be a female and what it means to be an engineer, you are more likely to be considering leaving your job, your organisation, or the engineering industry. This may mean you've spent time looking for another job, you often think about leaving (especially if things aren't going well at work), and you may have planned to quit the job, team, organisation, or the engineering industry in the next year.

These three direct paths, from identity interference to work outcomes, demonstrate the impact of what can happen when our gender and occupation identities conflict with each other in an engineering work setting. As the earlier statistics suggest, the effect of two identities interfering with each other, such as being a female engineer, can detrimentally impact individuals, teams, organisations, and the wider UK engineering industry.

17.9   INDIRECT PATHS – AUTHENTICITY AS A GO-BETWEEN FOR IDENTITY INTERFERENCE AND WORK OUTCOMES

Next, let's explore the indirect paths (arrows a and c in the diagram) between identity interference and work outcomes when authenticity is added as a mediator. This means we are looking at authenticity as a go-between for identity interference and work outcomes, with authenticity being caused by identity interference and authenticity impacting or influencing the work outcomes.

17.9.1 FAÇADES OF CONFORMITY

Facades of conformity, or the mask we can wear when our values don't align with the values of our organisation, worked as a mediator between identity interference and all three work outcomes: affective commitment, career compromise, and employee turnover intentions. This means that, as a female engineer, if you experience conflict between what it means to be a female and what it means to be an engineer, then you are likely to create a façade of conformity. When you create a façade of conformity, you may find this detrimentally impacts how you perceive your commitment to the workplace, you may feel that you are compromising on aspects of your career, or you may be considering leaving your organisation or the engineering industry.

17.9.2 PERCEIVED AUTHENTICITY

'Perceived authenticity' has worked as a mediator for identity interference on affective commitment and turnover intentions. This means that, as a female engineer, if you experience conflict between what it means for you to be female and what it means for you to be an engineer, then you are more likely to experience inauthenticity. This experience of inauthenticity may lead you to perceive reduced commitment to your workplace, or you may perceive you've been considering leaving your organisation or the engineering industry.

17.10 WHAT DOES THIS MEAN?

Quite simply, being female and an engineer in the UK engineering industry is, and is likely to remain, problematic unless we all act! Engineers are telling us that the interference or conflict they are experiencing at work can initiate façades of conformity and perceptions of inauthenticity, impacting their commitment, careers, and retention.

This feeling of wearing a mask or inauthenticity at work may lead to the feeling that their commitment to work has decreased, their career feels compromised, and their intention to leave their job, employer, or the engineering industry is heightened. The strong effects on career compromise seen across the three different time points, despite the smaller sample size in the last two surveys, indicate that reducing or eliminating the experience of identity interference and façade of conformity to increase authenticity for female engineers is critical to improving the work lives of our engineers and to gender diversity across the UK engineering industry.

17.11 SO, WHAT CAN I DO? THE THREE-STEP APPROACH

As a female engineer, know your values – explore them, define them, and review them using this three-step process. Your values sit at the heart of your authenticity. Your authenticity is critical to navigating your career and engineering. When you know your values, you must ensure that your words, decisions, and actions align with them.

> *I'll say that again: to be authentic, your words, decisions, and actions must align with your values.*

Set aside 30–60 minutes in the first instance, grab a notepad and a cup of tea, then follow these steps:

Step 1 – To explore your values, use an online toolkit, write a list of your dislikes and search for the opposite, or use a list of values and note the ones that stand out to you. Google is your friend; it holds many toolkits and lists – favour and use whatever works for you.

Step 2 – From the list you created in Step 1, define your three to five core values. Write these in a list, keep it in sight (on your phone, desktop, or a piece of paper on your monitor, journal, or fridge) and check it regularly.

Step 3 – Consider how you can best review your values and your authenticity journey, and schedule regular reviews. If your chosen values aren't working for you, or a significant event impacts your life, revisit Step 1. You could also ask your nearest and dearest to comment on either what they think matters to you or ask them to comment on your chosen values.

These three steps alone will help you at any stage of life or career, but don't forget to engage with your support network: your coach, mentor, manager, work colleagues, and allies. We all need someone who can help us to navigate our careers.

Suppose you are reading this as an *ally of women in engineering*; you will already know that the art of building and nurturing supportive relationships with underrepresented individuals or groups can be challenging. An allyship gap exists; this is the gap between the good intentions of the majority group and effective action as experienced by the minority group.

Whatever your position, be it female engineer, ally, having a lightbulb moment or still doubting my words' by understanding the experiences of your female engineering colleagues first-hand, you can challenge your assumptions. Ask your female colleagues about their experiences at work, with no judgement, without problem-solving, and without being the hero. Listen and note the blockers and barriers they face, especially those invisible to you, and ask how you can best be there to support your female colleagues. If asked, stand up to be their sponsor, coach, or mentor; everyone will want your allyship differently. Please support these amazing women; they are our mothers, daughters, sisters, and friends. They are the future of engineering.

17.12   WHAT ARE YOU WAITING FOR?

I hope you agree that engineering cannot wait until 2090 to have an equal gender ratio of employees. Diversity, not just of gender, brings positives to our workplaces, economy and all our lives.

We can't wait for someone else to act; everyone is responsible. So, what are you waiting for?

ABOUT THE AUTHOR

Helen Sara Johnson is a doctoral researcher (and possibly, by the time you read this chapter, a Doctor of Philosophy) from England, UK. She is also a human resources practitioner at the Manufacturing Technology Centre. Her academic accomplishments, BA (Hons), MA and LLM, align with a successful HR career spanning over 20 years. Inspired to give something back, she has presented her research through academia and industry, engaging with organisations including Jaguar Land Rover and the UK Catapult Network. Outside of academia, Helen is an autistic ADHDer, a champion of well-being and neurodiversity, has a passion for endurance running and is a proud cat mum. Helen's PhD thesis will be available through the Loughborough University Research Repository as and when completed. To engage with Helen to discuss her research, she can be contacted via Loughborough University, UK.

REFERENCES

1. https://www.wes.org.uk/about/what-we-do/resources/statistics/
2. https://raeng.org.uk/policy-and-resources/diversity-and-inclusion-research-and-resources/gender-pay-gap
3. Not all gaps are created equal: the true value of care work | Oxfam International. https://www.oxfam.org/en/not-all-gaps-are-created-equal-true-value-care-work
4. How advancing women's equality can add $12 trillion to global growth | McKinsey https://www.mckinsey.com/featured-insights/employment-and-growth/how-advancing-womens-equality-can-add-12-trillion-to-global-growth
5. Van den Bosch & Taris (2014). https://doi.org/10.1007/s10902-013-9413-3
6. Hewlin, P. (2003). https://doi.org/10.2307/30040752
7. Hewlin, P. (2009). https://doi.org/10.1037/a0015228
8. Kernis, M., & Goldman, B. (2006). https://doi.org/10.1016/S0065-2601(06)38006-9
9. Metin, U.B., Taris, T., Peeters, M., & van Beek, R. (2015). https://doi.org/10.1108/JMP-03-2014-0087
10. Settles, I. (2004). https://doi.org/10.1177/0146167203261885
11. Meyer, J., & Allen, N. (2004). https://www.employeecommitment.com/TCM-Employee-Commitment-Survey-Academic-Package-2004.pdf
12. Creed, P., & Gagliardi, R. (2015). https://doi.org/10.1177/1069072714523082
13. Rusbult, C., Farrell, D., Rogers, G., & Mainous, A. (1988). https://doi.org/10.2307/256461

18 Allow Them to Answer the Call

Shane Woods
Girlstart, USA

"Can I ask you a question?"

A common question I am asked by caregivers of kids showing an inclination towards building, creating, or problem solving is, "How do I support my child who loves STEM?" This question is typically followed by personal comments that take the child out of the center and replace it with the nostalgic adult, such as:

- I was never good at math/science.
- I always hated math.
- My science teacher in fifth grade was awful.
- And so on, and so on.

I quickly let them know that it is important to change that language so that it doesn't discourage their children. Contrary to popular beliefs, kids do in fact listen to the adults in their lives. I say this as the parent of a child who at the age of 4 yelled a well-placed "Damn!" when she dropped a toy in her Christian pre-K class on grandparents day. I certainly learned my lesson that day while collecting the best parent fail of the day award. Kids are ALWAYS listening and it is imperative that we make sure their environment is filled with words of promise, words that encourage risk taking, and most importantly, words of support.

Anyway, back to encouraging humans who are experiencing childhood as girls to take STEM-aligned courses in secondary school and beyond. For all of us who hold the formidable positions that require us to offer guidance to adults in charge of someone in the present-day K-12 system, we realize that our job of transforming a fixed mindset to one that embraces growth lies not only with children, but also with those they spend most of the time with when not in a classroom or out-of-school STEM experience. These are caregivers, parents, guardians, even neighbors. However you identify, please know that it is to you that I direct the rest of this chapter.

Your past experiences in mathematics and science courses or in school in general should not inhibit you from providing the environment your kid needs in order to flourish in STEM competitions, classes or majors. I intend to provide suggestions that can be implemented this week so that it becomes muscle memory, so you can get out of the way and allow your child to answer the call of STEM.

DOI: 10.1201/9781032679518-21

18.1   THE CHALLENGES OUR KIDS ARE FACING

Our children are up against three challenges that need our attention. One challenge is that our youngest learners and those who live in under-resourced communities are sitting in classrooms receiving insufficient science instruction. Science time, where most introductions to STEM are found, is being ignored or shortened in order to redirect all resources to reading and mathematics in elementary. The 2019 NAEP Science results (Nation's Report Card 2019), show that across the grades, few students exhibit science proficiency, which means that by the time they reach 12th grade, only 59% scored *above basic* in science. One way we can correct this issue is by championing for more science minutes during elementary school so that our kids have the foundation needed to take on more rigorous science and math courses in middle school and beyond.

The second challenge falls on teacher preparation programs that don't adequately equip them to provide quality science or STEM education. Elementary teachers typically feel well prepared to teach reading and mathematics. Secondary teachers may have a generalist certification that does not provide technical training in the more advanced science courses. Professional development opportunities may be limited or non-existent depending on how the school district is structured. Enrichment in science education may become the responsibility of the teacher and many have to prioritize the initiatives being required by their building or district administrators, meaning that science may be low on the totem pole.

The final challenge meeting our kids at the school is the astounding reality that quality science instruction is not equitably distributed. We have all seen movies that show the reality of an undertrained teacher being assigned to the kids who demand the most from an educator. Of course, we have seen the success stories like those of Jamie Escalante in *Stand and Deliver* or that of Erin Gruwell in *Freedom Writers*. Unfortunately, those resilient teachers are few and far between.

The reality is that most new teachers leave the profession within the first few years, and after Covid, that statistic has worsened. The resources available to the teachers can vary from building to building. The availability of subject-specific clubs and competitions may not exist, leading to diminished opportunities for developing an interest in STEM courses. There is also the sad reality that not all high schools offer advanced science and math courses due to a lack of qualified teachers.

Our kids are up against formidable challenges, and it will take the adults to find the root cause and dismantle the system that feeds the problem. You are already taking the first step by reading this book and garnering the wisdom within in order to create your own action plan. It is going to take us all to change education for the better. We saw the levels of innovation which are possible, as we all witnessed education pivot during the 2020 shutdown. We can and must answer the call to make education work for our girls.

18.2   WHY IS IT IMPORTANT TO MAKE STEM ACCESSIBLE TO GIRLS?

It is widely supported that girls' and young women's achievements are on par with those of boys and young men when it comes to science and mathematics. They are taking advanced courses in mathematics and science, but they seem to be lagging behind in enrolling in advanced physics and computer science courses. As a former district science leader, the first set of questions that come to mind are,

> Why is that? Are counselors not advising kids based on potential success in these courses? Are teachers having the inspirational conversations to entice a girl to enroll in an advanced class? Have they had a chance to see a woman that looks like them in a career connected to those courses?

Too often, the answer to all of these questions is a resounding NO.

As the adults with the power to change the systems as they currently exist, we need to focus on increasing access to STEM for girls, making sure K-12 teachers, college faculty and informal educators are not just aware but proficient in using key tactics that nurture the mind of girls as well as craft STEM experiences that allow them to build their skills in math, science and engineering practices. Making sure that the youth have access to opportunities that are high-quality, relevant and inclusive can happen in and out of the classroom.

Increasing awareness also means catching the eye of the caregiver who makes the final decision on whether the kid can enroll in that class or attend the out-of-school program. If it is not happening within the school day, then price, location, day of the week, time of day, length of commitment and trust within the community are all factors which are considered when making the decision to sign a kid up for a STEM program.

After nearly 25 years of working with girls, I have learned that they are most interested in learning about careers that impact their world, careers that make a difference, careers that allow them to leave a legacy. The programs I have led and continue to influence hone in on the strengths, needs and challenges of girls so that they feel like they belong in STEM. Girls must have a menu of choices and opportunities to encounter STEM in all its forms in order to find the path that intrigues them.

At Girlstart, a nationally recognized informal STEM education program, we focus on sustaining an ecosystem that thrives on peer and near-peer support as well as positive youth development by offering relevant curriculum through supportive relationships with adults that include guest speakers and adult volunteers. Our free programs allow girls to work side by side on STEM challenges led by staff who are typically enrolled in a STEM major at a local college or university. The curriculum used always includes a woman with a STEM career based on the lesson they are learning.

We make sure they are aware of the massive support they have from nonfamilial adults in the community when they are working with volunteers to succeed in finding a solution to an engineering challenge. Guest speakers and volunteers embody the careers that are changing the world and make it real so that the girls know they too can achieve what may have felt impossible.

Studies show that gender differences in attitudes and interest in science are present by the end of the elementary grades. These early years, therefore, represent a crucial window not only for providing students with a solid foundation in STEM subjects, but also for cultivating an enthusiasm for STEM among girls.

18.3   THE ROLE OF CURIOSITY

Kids are naturally curious. They are scientists before they even learn the word and definition. They test out hypotheses, conduct trials to support or dismiss their claims by actively collecting data and, ultimately, communicate their success or calculated failure. If you are still unsure of this happening, let me give you this example.

Have you ever handed an unschooled child an object which they are unfamiliar with? I have, and this is a non-scientific observational account of what I saw. I handed my three-year-old nephew his birthday gift in a wrapped box. His instinct to rely on his senses immediately kicked in. He shook it. He turned it over and over in his hands. He smelled it. Of course, he even resorted to asking what was inside. He provided a few guesses based upon what he figured could fit in a box that size. When his patience was exhausted, he finally took the paper off and looked inside. He was pleasantly surprised to see a battling spinning toy he had asked for. He was using estimation without formal training. He was inferring before he had one science class. All he had at that age was curiosity, and it led him to model behaviors used by scientists and engineers around the world.

Caregivers also tend to ask me how they will know that their child is ready for STEM. I always say they are born with the most essential skill and behavior, *curiosity*. We can help them refine and harness that curiosity by immersing them in environments that reinforce that academic abilities grow stronger over time and show them girls and women who have succeeded in math and science. The goal is to foster a long-lasting interest in mathematics and science founded in curiosity about the world they live in, which will allow them to persevere even when the material becomes challenging.

As you seek out opportunities for the youth under your care, create a list that you feel will support curiosity in a STEM setting. The following guiding questions are a small sample of criteria you should consider when evaluating a STEM program or course of study that will foster their growth in STEM.

- Will my child apply math and science as they consider real-world solutions to a science or engineering problem?
- Will they have the opportunity to hear from experts in the field they are studying? Will the adult volunteers resemble the girls?
- Will they work in teams and be encouraged to engage in safe and healthy dialogue even when disagreeing?

If no class or after school programs meet these minimal requirements, reach out to your school's counselor or attend a board meeting to ask about the process of adding these types of courses to your child's school. You can also seek out national STEM programs that may be able to come to your community to meet the needs of girls who have the potential to excel in STEM. A great place to start is the National Girls Collaborative Project (ngcproject.org).

18.4   HAVING THE CONVERSATION

As I walked with my daughter from one section to the next of the world-renowned theme park we all dream to visit at least once, I stopped and took a picture of a sign that spoke to me. It happens to belong to the doctor who helped us rhyme and even possibly led many of us to the misconception that we could make millions as a rapper, but yet again, I digress. The quote was from the one and only, Dr. Seuss. The sign read:

> Think left and think right and think low and think high. Oh, the things you can think up if only you try!

This simple message sums up what I have always wanted to instill in all of my students, campers and after-school attendees. Resilience looks different for every individual. For the oldest child acting as a surrogate parent for their siblings while their mother works multiple jobs to keep them housed, their capacity for resilience cannot be compared to that of an only child born to privilege and position.

However, both need to know it is okay to keep trying when they encounter something new and possibly challenging in school. They need to know it is okay to take healthy academic risks and bet on themselves. By introducing them to STEM challenges, educators are giving them the ability to experience stress and find solutions in a group setting, which is encountered just about every day in the workforce.

As a caregiver, take the time to have conversations with your child that go beyond asking how their day went. Ask if they did group work and how they contributed to the project. Ask if they found or listened to multiple solutions to a problem the teacher assigned. Ask if they learned about a new career, and whether it was a woman or a gender-expansive person they identified with. These types of questions will allow you to gauge the quality of instruction they are receiving and to determine if they need more advanced experiences.

While talking to our kids, we must teach our youth to speak to themselves, using encouragement and healthy optimism that allow them to course-correct without wallowing in failure as something that cannot be overcome. Ask them to share how they learn from a failure. Ask them what they say to themselves to motivate them to not quit. We must reinforce the importance of developing a positive self-identity because there will be times when the only motivating voice they have is their own.

18.5   HOW CHEER CAN INFLUENCE US

Beyond the expertise I gained from teaching middle school science, leading a district of 88,000 K-12 students in a Texas district as a science director, reimagining a 90-acre camp to build STEM interest, confidence and identity, and acting as an executive director for a nonprofit that empowers girls to stay curious about STEM through year-round out-of-school programming, I can tell you that everything I shared with you so far works because I am the parent of a living success story.

My daughter triumphed at every level of her matriculation through K-12 and college despite the historical exclusionary practices put in place to discourage those showing up at the intersection of gender, race, ethnicity and class. Knowing that Black, Indigenous and Latina women represent less than 10% of the STEM workforce, my daughter is left-handedly adding to that number by pursuing a career in human-centered design and engineering. She answered the call to become a woman in tech.

She was able to do that because of her own beliefs, skills, educational experiences and confidence that was amplified by having a support system that identified her as belonging to and in STEM spaces. I was intentional about feeding her curiosity in STEM by seeking out after-school and summer programs that encouraged her to learn more about careers associated with math, technology, science and engineering. As she grew older, her belief that she could do STEM led to her applying to an accelerated biomedical program at a neighboring high school.

Once accepted into the prestigious program, she used her scientific skills to work on real-world scenarios and excelled in her math and science courses. She even earned college credits as a result of her AP exam scores in biology and psychology. Because of her academic achievements and out-of-school participation in STEM-related activities, my daughter recognized the value of STEM and how it plays a role in everyday life.

I have seen how it matters to show up for your child as they test-drive new interests. Seeing your child perform the audience scan looking for a family member and then locking eyes with you and smiling or waving or blushing is a feeling like no other. She did that way back in 2006 at her first school assembly when she had a solo, and she repeated the behavior in 2023 when she walked into the auditorium at Texas Women's University to receive her bachelor's degree.

You, the caregiver, are showing up as their personal cheering section. Our kids who are brave enough to try STEM need a committed cheering section, and it must show up when helping them work on a science fair project, it is needed when they enroll in AP courses and silently plead for a quick pep talk before walking into the testing site and it is critical on that day, they receive the well-deserved recognition of Latin Honors at graduation.

Our presence confirms that all of the hard work was worth it and it allows them to be proud of themselves and to become comfortable winning in life. It is imperative that they enjoy those wins because they will continue to show up, and they must learn to dismiss the message of the naysayers who speak of humbleness. No more dimming the light of our up-and-coming STEM leaders. We must teach them to cheer by modeling what that sounds like and feels like.

I sincerely hope that by contributing my perspective on how those of us who have been granted the ability to walk this land longer than the youth we love and nurture can become the adults so many of us grown women in STEM needed when we were girls. It is critical that we remove the roadblocks we encountered or witnessed others navigate as we pursued STEM courses in high school and college. The ultimate goal is to guide, mentor and sponsor the next generation of women in STEM.

We need them to know they belong in those STEM classes, camps, competitions and careers so that they stay and thrive in the field. If you find yourself face to face with a young person with a healthy dose of curiosity and an inclination to numbers, by all means, do whatever you can to slingshot them into the rewarding and remarkable STEM universe. Our future depends on them.

ABOUT THE AUTHOR

Shane Woods began her career as a middle school science teacher in the Fort Worth Independent School District. In her 17 years with the district, Shane was able to make her mark in every position she held, from department chair to leading the entire district as the K-12 Science Director overseeing curriculum and assessment development, while supporting teachers in honing their skills through year-long professional learning opportunities. From 2019 to 2022, Shane acted as Senior Director of the Girl Scouts of Northeast Texas STEM Center of Excellence, which is a 92-acre living laboratory where kids of all ages can explore and develop their competence and confidence in science, technology, engineering and math – all while cultivating essential skills such as confidence, resilience, leadership, risk taking and problem solving. In her newest role, Shane serves as Executive Director for Girlstart, a nonprofit focused on empowering girls in science, technology, engineering and math. She remains active in the formal and informal education community as a highly-sought-after nationally recognized and award-winning leader and speaker in science education, equity in STEM and leadership. She has made appearances on several podcasts such as *Quietly Visible* and enjoys teaching as a part of a panel for nonprofits like Play Like a Girl. As a communicator who immediately establishes a rapport with those in attendance, Shane can speak to any size group from a small staff on Zoom to a room full of statewide science leaders. Shane also gives back to the community through the leadership positions she holds in organizations that champion youth development and education such as the National Science Education Leadership Association, the American Camp Association and the Texas Girls Collaborative Project.

REFERENCE

National Center for Education Statistics. (2019). *Nation's report card*. National Assessment of Educational Progress.

19 Through It All, the Why Was My Motivator

Claudia Gomez-Villeneuve
MacEwan University, Canada

On my first day in engineering school, the ocean breeze was warm, the parakeets hiding in the palm trees were chirping, and the air smelled of sweet cayena flowers. Why? Because my first day in engineering school was in Colombia, South America, at a beautiful university near the Caribbean Sea.

The decision to study engineering was easy. First, my parents were very clear that I could study whatever I wanted if it was related to science and technology – that narrowed the field down indeed. Second, I had two life-changing and inspiring high school teachers, one who taught physics, and another who taught chemistry, so I was set. And third, and last, I attended career day at the aforementioned university-by-the-sea and the speaker who explained Industrial Engineering spoke directly to my 'Project Manager' soul – by saying I could manage teams and guide them. I felt I just had to, had to, study industrial engineering.

Fast-forward 12 months when I had just finished my first year of engineering school, and my family received the news that our immigration visas were approved to move to Canada. For the next few months, it was a whirlwind of activity to pack the essentials for a family of five and say goodbye to our life in Colombia. Of course, I had zero doubt that I was going to resume engineering school in Canada, so the only question was: when I would be able to start studying again.

19.1 MOVING TO CANADA TO STUDY ENGINEERING

After arriving in Canada, it took over two years to finally resume engineering studies at my new university-by-the-North-Pole. This is an exaggeration, of course; it wasn't quite the North Pole, but it was darn close. It was a university in Edmonton, Canada. My only barrier to entering school was to upgrade my Physics 30 mark, because the admission team rejected it. So, I enrolled in high school classes, studied my little heart out, and passed the exam with a better mark than before. The whole English language was not really a barrier, since I studied in a bilingual primary school before, but it was certainly another fence I had to jump. I submitted the new physics grade and waited for a response from the university. And I waited, and waited some more. Eventually, I called them directly because classes were going to start soon, and I had not received a response.

While on the phone, I was told by the admissions officer that my physics grade was just over the threshold for acceptance but not high enough to impress them, so

DOI: 10.1201/9781032679518-22

their conclusion was that I might find engineering school too hard to survive. That was all the motivation I needed, for them to think I was not good enough. I replied in my much-improved English that I had been valedictorian at my high school in Colombia, that I was a hard worker, and that I was going to be an amazing engineer; and added: so of course, they had to let me in. It worked. The male voice on the phone said, "Okay, you are in!" Is that what the officer needed to hear? That I was confident? Yes, indeed.

Fast forward six years, and in my hands now I had a shiny frame with my bachelor degree in civil engineering (the university-by-the-North-Pole did not offer industrial engineering, but it was close enough to project management, plus I love bridges and structures); and also a shiny frame with my master's degree in construction management. Staying in school a little longer for my master's degree was recommended by my favourite engineering professor, a woman, my only female instructor for six years, who knew I wanted to specialize in project management.

I had also learned from a summer job that engineers spend a lot of time managing projects and communicating with colleagues and clients; therefore, I needed more training on that, beyond being an expert in math and calculus. This was the reason I took the master's degree – to learn project management. Then, to finish it as quickly as possible (after all, I had a baby and a husband to support), I chose the course-based version as opposed to the thesis-based version, and also compressed all my graduate courses into one calendar year. No one in the program had ever done that before, but after I did it, many fellow students followed my lead.

19.2 GRADUATING AFTER MARRIAGE AND A BABY

After six years in Canadian engineering school, I had a shiny new husband, and a shiny new baby. Yes, I fell in love with a Canadian boy who looked just like Tom Selleck from the TV show *Friends*, so we got married after my first year of engineering and then we got pregnant while I was finishing third year. Talk about multitasking – the office secretary even said she had never seen a pregnant engineering student before, and it makes sense, given that there were so few women who are engineering students. So, I had my beloved engineering degree (two of them) and a husband with a baby. All I needed now was a cool job.

One day at the student job office in the university campus, I saw a job posting looking for a "Project Engineer-in-Training" to work for a pipeline company, and the description was perfect for me. Thankfully, when I applied for the job, I already had a master's degree in engineering which easily beat all the other applicants who had, so far, only completed a bachelor's degree in engineering.

I was lucky that I had job interviews with three different companies, and they were all so much fun, but the pipeline company was by far my favourite job interview. I prayed for them to call me, but of course, they took the longest to get back to me. I even turned down the job offers from the other two companies, praying that I was going to get a call back from the third. My prayers worked. I started working as a project engineer in training for a big company, just a few weeks after completing my master's degree and when my son was almost three years old.

The company was huge, with thousands of employees, but it felt like family. The working culture was friendly, and the work was plentiful. My bachelor's degree and my master's degree were immediately put to good use, and I began the hard climb up the corporate ladder. It is important to note that in the 16 years I worked at this pipe-line company, all my engineering bosses were men; that is how the cookie crumbles still. As you will read in this chapter, some of my bosses were supportive while others were just clueless.

My second baby was born six years after the first baby, and I mistakenly thought that this was not going to affect my career much. Boy, was I wrong. As soon as my clueless boss found out I was pregnant, the questions began: "When are you going on maternity leave?", "Do you want us to give your projects away?", "Are you ever coming back?".

Remember when I shared before that I had been valedictorian at my high school in Colombia? Well, that was to show you that I was no stranger to hard work and discipline. I used that hard work and discipline to complete my engineering school in a new country, Canada, and in a new language, English. Now, all these questions were making me feel confused, as if my engineering life was somehow over, fin-ished, barely two years after graduation. Again, that was all the motivation I needed, for them to think I was not good enough. I went on maternity leave, determined to come back and succeed again!

The second baby was a girl, and the story of her birth still motivates me today when I think life is too hard or when I imagine there are too many barriers to make my dreams come true. Her birth was so amazing, it still fuels my motivation to ask for what I really want. Actually, the story of my second baby's birth starts with the birth of my first one, six years before. Like many first-time mothers, I had foolishly thought that all I had to do was show up at the hospital and the doctors and nurses would take care of me and the baby. There is so much more than that. The legal and ethical research field of "patient rights" is gigantic.

After the birth of my first child, I realized that hospitals are a business and that we should treat doctors and nurses not as our friends or caregivers, but as fellow professionals. They offer a certain list of services and we, as clients, must decide if the services are acceptable or if we need a second opinion or even a third opinion. I went with the only opinion I asked for and as a result, the pressures of the hospital operation, their traditions and fear of medical liability risks, left me highly drugged and surgically bandaged, and holding a new baby who was also highly drugged and surgically bandaged. Everyone said that was a normal birth these days. I should have asked for a second or third or fourth opinion. I stored that pain inside me for six years, and it all came forward in a torrent when I discovered I was pregnant again. It opened the floodgates of my past, and I swore things would be different this time.

My second birth was completely different, not just in the mechanics of it but in how I communicated with the medical establishment. After months of research, my daughter was born via a homebirth in a small pool of warm water, attended by a pro-fessional midwife and also by a certified doula. I used all my engineering problem-solving training, my ability to research journals and understand statistics, and I got the most natural and surgery-free birth I could ever get. After 12 blissful months of paid maternity leave (a fact in Canada), it was time to go back to climbing the corporate ladder.

19.3   ASKING FOR PART-TIME WORK AFTER MATERNITY LEAVE

Returning to work in engineering after a long maternity leave was no picnic, not easy at all. A few months after I began maternity leave, my boss called me at home to let me know my position as project manager was being moved to a different department. When I arrived at the office, I realized the new job role was more of a demotion, since I was not a project engineer anymore; now I was something called a compliance engineer. The job tasks were much simpler, and my level of power and control had been severely reduced. I had been warned, by fellow mothers in the office, that this was a risk of pregnancy: job demotions. The salary pay may not have changed, but the type of work had been reduced in importance. To complicate matters, this new job was not technical enough to truly qualify as an "engineering" job.

Since I had just graduated from engineering recently, I was still considered an EIT or Engineer-In-Training; in other words, I was still on 'probation'. In order to complete my training, and eventually become a professional engineer (P.Eng) myself, I had to continue performing engineering work under the supervision of another professional engineer for another two years until I qualified. Therefore, the new job as compliance engineer due to my pregnancy and maternity leave, effectively put the brakes on my career growth. I reviewed the P.Eng. license requirements and only 10% of my new role met the requirements. At this rate, I calculated it would take me about 10 times as long to be promoted to the professional engineer role. I did not like that estimate, especially after I put so much effort studying and working. Plus, now I had two children and one husband to feed, so my continuously growing salary and my career had to climb up, not go down into a hole.

I took as many "stretch" assignments (tasks beyond your role, to learn new skills) as I could, with the approval of my new boss. Because my ears were perked up, I heard about these stretch assignments everywhere: during client meetings, staff meetings, and even at the watercooler. (Well, it was not really a watercooler, but it was next to it, in the coffee room.) These additional tasks beyond my role achieved what my actual role did not, I was able to submit the paperwork and eventually qualify for the promotion to professional engineer.

Let me explain the part-time job status now, because this is an important part in the story of my life as a female engineer. After the birth of baby #2, after that glorious waterbirth, I was full of confidence and ready to continue asking for what I wanted. About two months before going back to work after maternity leave, I emailed my new boss asking for a part-time job. I had done my research and I had used my project management skills to plan my new life as an engineer, a wife and a now mother of two children. I tried different arrangements and it came down to one simple conclusion: I needed to work 80% of a full-time schedule to make it all fit perfectly.

I sent the email requesting a 0.8*FTE, or 80% of a Full Time Equivalent, and because my new boss was an accountant (not an engineer – remember I had said my job after baby #2 was a demotion), and since accountants are more familiar with part-time work, he said yes. So the demotion was a blessing in disguise. Thank you! I don't know why engineers don't know about part-time work, while accountants and other professionals seem more knowledgeable, but my part-time status was accepted.

I became the poster child for this because many women engineers in the company followed my lead. I believe now that when my accountant boss approved my 80% FTE, he did a great favour not just for me, but for the company, and for the engineering industry in general. This approval created a "legal" precedent (like lawyers, where they use a previous legal decision to support future decisions) that allowed fellow engineers in my life, of various genders, to think hard about how many hours they were putting in at work. All they had to do was sacrifice 20% of their salary, like I did; it hurts at the beginning but eventually you forget. Now that I am older, I see why people begin to value time over money, but back then, I made the request just to keep my sanity and balance my work with my life. I worked as an engineer for almost 16 years in that company, and the last 12 years of it, I was a part-time engineer.

19.4 OVERCOMING A MAJOR CAREER OBSTACLE

Unbeknownst to me at the time, when I achieved the P.Eng. just five years after graduation, I had broken not just one but two barriers that block the successful engineering career of a woman in the world. The two barriers I broke were: staying in the career past the first five years after graduation; and obtaining the professional license. That was a critical moment in my career, I just did not fully understand how close I came to failing.

I met many women who were successful in their careers even after they left the field of engineering, but I just did not think I wanted to be one of them. I wanted to be successful in engineering, and by going back to "WHY" I was doing all this, because I enjoyed the work and I enjoyed the power it gave me. The salary pay was not bad at all, I was getting paid about the same as other men in engineering, which is a gift because I was told women are normally underpaid for doing the same work men do. So, I loved my work and I loved the benefits; therefore, staying in engineering was definitely the goal. Then, I got pregnant again.

My third child came 6 years after baby #2, and 12 years after baby #1, so I knew I had tried to keep my career in place, which is why it was excellent that the pregnancies were so far apart. I was thrilled to have another baby, but I was no fool, and I knew that the impact to the career I wanted to maintain could be disastrous. I knew when my career would be affected, I just did not know exactly how. My baby #3 came out as beautiful and sweet as I predicted, and leaving him in daycare 12 months after his birth, at the end of my maternity leave, was extremely painful. But this was nothing compared to the pain of getting my engineering job back.

I had worked for the same big company now for almost 10 years, but office politics and reorganizations are the same everywhere; when it was time to return from maternity leave, I found myself without a position to come back into. My job role had simply disappeared and the only instructions from the Human Resource Department were: "You still have a job here (as per the law required) but you need to apply for any jobs available by yourself." I was scared to be fired, so I applied for many internal jobs in the company, and I also did strong networking. I emailed my past bosses and my colleagues at the office telling them I was available to transfer to their department. And since my reputation preceded me, I was able to find a job.

One of my past bosses, from my first year at the company, gave me that job. I would have been thankful for that new role, even if I had to become a coffee engineer, an office assistant, because it meant I still had a job. But the role he gave me was the coolest job ever, project engineer and project manager, one that finally brought my career back on track after many years. It was the perfect job to match my education, and the perfect job to match my professional aspirations. Now, I could continue climbing the career ladder with the proper credentials.

19.5   REALIZING MY SALARY NEEDED A BUMP

At a certain point in your life, you feel old enough or smart enough – take your pick – to become a boss at work, to hope for a career in the management track. Like many female engineers, I had seen male engineers my age, my past university classmates, climb into management many years before me. These men also had children, but their babies had not affected their career negatively; in fact, some of them seemed to gain greater respectability after having children. These new fathers now looked like management material. Not so much for the new moms. So I had started to notice that the majority of female managers, directors or vice-presidents did not have children younger than 15 years of age, or they had no children at all. With my three children, a supportive husband, and a decade of engineering work under my belt, I felt ready to 'lean-in' as the book advises. So, I leaned. And I leaned hard.

It took a few tries, and new skills had to be developed, but the oddest thing was when I discovered that even my salary needed to be upgraded before I could be considered for a management role. How is that, you say? How does your salary level become a factor in a management promotion? Well, my female salary at the start of my career may have been the same as a male salary, at the start of their career, since we are both in engineering. But as years passed, my salary did not climb as fast as the salary of my male colleagues did while doing the exact same work. I confirmed this when I opened a local magazine who had done a salary survey for companies doing engineering work. Reading the magazine, I had no idea I was about to discover that I was underpaid. Not a little bit underpaid, but severely underpaid!

The magazine had detailed tables showing annual salary by type of engineer, type of industry and level of authority. It took me a few minutes, but I found myself on the table. The average annual salary for a person doing my role exactly – that is, the majority of people executing that work – was 30% higher than my annual salary. Even after adjusting for my part-time status (more on that later), I was severely underpaid. It felt like a knife to the heart. I always thought my pay was on par with the industry, on par with the other people in my department, but I was very wrong.

Somewhere in the last 10 years, after two maternity leaves, one job demotion and not a single female boss, I was at the bottom of the pay scale for the work I was doing. Thankfully, I am a person that sees barriers as problems with solutions, so I decided to solve this one. One of my many un-official mentors had pointed out that if I wanted to get a promotion to management, my salary had to be at least in the vicinity of the entry level of the management pay scale. If my salary was too low, the promotion could not be processed from an accounting restrictions perspective. So I planned to fix my salary to get at least the average, which meant I had to talk to my boss. I prepared my speech (wrote it, actually – 200 words), printed my updated resume and brought the salary survey magazine with me.

He listened calmly and had a few questions. Then he fully agreed with my analysis and promised to do everything in his power to bring my pay up. He did another great thing that day: he apologized for not noticing this gap in my salary. But the best thing he did came later, when he actually submitted the request to Human Resources, to adjust my salary. My mentor was right – there were accounting restrictions to increasing someone's salary suddenly. My boss had to fix the approximately 30% gap in my salary into two adjustments: a portion in the first year, and the rest in the second year. Within two years of my request, though, my salary had become the average for my job band or job level. Now, I could finally hope to be considered for a promotion to management.

19.6   GETTING A PROMOTION TO BOSS

Now that I was back to climbing the corporate ladder, I started to notice my colleagues in engineering were advancing in their careers faster than me, especially the male colleagues my age. I had heard they were being asked directly to apply for management roles – the good old "tap on the shoulder" – but I was not getting asked. Also, I noticed that even when I brought up that I could be interested in a management role, that there was no enthusiasm from other colleagues.

I really wish people at work were more open about sharing how much the company pays them, or more open about how they get promotions at work. I think it is because of the secrecy that companies can hide how their pay is unequal between men, women and other genders, even when they are doing the exact same job; and they can hide how their rate of promotions between men and women is so unequal. I really think, now that I am older, that conscious bias or discrimination against women is keeping both our income and our rate of promotions lower than men. With all the training we have now on EDI – equity, diversity and inclusion – and the understanding that even unconscious bias or discrimination (the bias we are not aware of) is hurting women's pay and promotions in engineering, and the rest of STEM, it is time that salaries no longer remain a secret.

The antiquated notion that women don't need the extra money because they are not the primary breadwinners has to disappear in a society with high divorce rates, parental care arrangements and many diverse family units. Fair pay is actually about human rights. If salary ranges for each job role in a private company were published, like they are published for government or union jobs, it would be impossible to hide the conscious or unconscious bias anymore. Managers would have to explain publicly why they are paying some people more than others for the same work. I was so pleased now that my pay was fixed, because now I could focus on the other problem: when was I going to get promoted at work?

I had noticed for a few years how the lack of advancement in my career, meaning being given larger projects, being given greater responsibility or being assigned a team to supervise, was holding my career back. I was now marking 10 years after graduation, and I was not moving up. So, just like I fixed my salary, I resolved to fix my job title. I knew now that my salary growth depended on my position growth, but I was scared to become a boss myself. I had this notion that bosses had to be mean and angry to get any work done, and I did not want to do that. But it all changed when I heard a guest speaker at one of the company's networking lunches for women in engineering.

The speaker said that to be a successful boss, we had to be tough on the problems and work hard to solve those problems, but we had to be gentle on the people, since they are the secret to success. With that mantra, *"tough on the problem but gentle on the people"*, I had everything I needed to go find a promotion at work. So, I went to see the boss of my boss to find it. Note that it makes sense to go over your boss's head for this one because you are trying to get a promotion, and only your boss' boss can grant that.

Again, I was nervous to call the boss of my boss to set up an appointment, but he was also open to hearing what I had to say. I was thankful to the universe for giving me so many open-minded and kind-hearted bosses. Unfortunately, over the years, I had as many enlightened male bosses as I did scary male bosses. These scary bosses had a twisted sense of the role of women in society, and either they plainly ignored me or they fully threatened me. It is a blessing that I worked for a very large company, because I could either wait for them to be fired or I could move to another department. But my current boss was great, and his boss was even greater.

This time, I showed up at his office and told him I was interested in switching from a technical role to a managerial role. He listened calmly, asked me a few questions and reacted the same way I did: he was floored that I was not a supervisor yet based on all my experience with the company. He said he would keep me in mind for future open roles in management, and suggested I take more "stretch" assignments in the meantime. He said a "stretch" assignment, or a big task outside of my normal role, would raise my profile in the department because right now I was very invisible. That is, he would have to justify my hiring as supervisor to his own director and fellow managers, but they needed to know who I was. If my profile was higher, then he would get more support for a future promotion.

That is what I did. I began to ask around, and I found a new committee that needed a chair. I took the additional role, combined with my regular work, and it did expose me to new people in the department and to new challenges. Within two years, my boss' boss told me he had a management role for me. I was to be a supervisor of eight staff. I was so happy! My promotion came with a salary bump too, oh yeah. But the good news was short-lived, because my boss called me while I was in the car waiting for my kids after school to explain there was a problem with my promotion. He said that he was processing the promotion but that I had to change my work status from part-time (32 hours a week) to full time (40+ hours a week) because no one in management had ever worked part-time before, so he was getting resistance.

It helped that the phone call came when I was outside of my children's school, because I could see clearly what I would have to give up to become a supervisor. I stood my ground. I said I still wanted the promotion but that I wanted to keep my part-time status. My future boss chuckled on the phone and said, "You are so honest, at least you work 32 hours a week and we pay you for 32 hours a week. I have full-time employees who also work only 32 hours a week but I have to pay them 40 hours a week. The promotion is yours, working part-time, but we will review it every year." I was jubilant again!

My years as a supervisor were a dream: (1) I had the job I always wanted to have, to help and guide my team; (2) I had the cool title of Supervisor of Construction Projects; (3) I was still working part-time, so my family still had a part of me; and (4) my pay was fully synchronized to my role and my years after graduation. I was getting closer to the roles my male peers (in age and experience) were getting. It was a new world of fairness and equity. I can say confidently that I saw the future, the future of EDI, where people get salary raises and salary promotions just because they are awesome regardless of their gender or family status. I was so thrilled to be one of the first women to get promoted to supervise construction projects in our company. Thrilled and also sad, because I was one of a very few women. I wore that badge with honour and tried my best to support other women who wanted to get a promotion.

19.7   LEAVING MY JOB AND CONSIDERING QUITTING ENGINEERING

Being part of management was definitely a new experience for me, by being able to see "behind the curtain" at all the things management does. It was mostly good to see the secret stuff, but it also had its horrendous moments. The worst one was when the oil and gas market collapsed and I had to start firing people and cancelling projects. After three years in the role of supervisor, the price of oil crashed so badly and for multiple years that my job changed, from hiring new personnel and guiding them, to having to choose who to fire or let go, and in what order. The market crash created two factions at the office: those who fire people, and those who get fired. It was like being at war, and the more people we fired, the more horrible the situation became.

Eventually, workers did not know what was worse, emotionally speaking: to be laid off from their job, or to keep their job in a company where the employee culture was collapsing. And when you are in management, that dilemma is made worse because you are the person firing them, and you know weeks ahead that they are getting fired, but you have to lie all the time. So that is how I left my long-time employer, where I had been working for almost 16 years, and a company that felt like a family. I had the life plan to work there until retirement, but life is full of unexpected changes, so I left the moment they offered me the option of a lay-off package.

Honestly, it took leaving the company to realize how much stress I had accumulated, a stress that I was not doing a very good job at hiding. I specifically remember two things from that time: first, having an anxiety attack a few months before I left the company, and, second, taking only two seconds to make the decision to leave the company. The transition from overwhelmed employee to doing nothing was so rough, it brought the realization that perhaps engineering was not the field for me anymore. The complex work, and being a gender minority suffering the consequences of conscious or unconscious bias by the majority gender, made doing engineering for another 20 or 30 years an unmanageable option. Women are burning out managing those two worlds: work and politics.

Then I read an article about women in STEM – Science, Technology, Engineering and Math – and there in the article they listed how difficult it is for women to stay in their STEM careers for life because of two main crunch points: either after 5 years or after 15 years. The article described me perfectly: high achiever, wants it all, but either leaves STEM in the first 5 years after graduation when children arrive (I asked to go part-time exactly 4 years after graduation because my life became harder to manage); or leaves 15 years after graduation after achieving a certain level of success (I left the company after 15 years and considered leaving the field). What? The article described me perfectly.

The first time I felt the crunch, my nice accountant boss accepted my request to work part-time, so I continued to work as an engineer. The second time I felt the crunch, I left the company with a nice lay-off package but seriously considered leaving engineering. This was a critical moment in my career; I had to make a decision. Then, I remembered why I became an engineer: I wanted to help the world, and I wanted to build fabulous teams and great systems and structures. "If I leave engineering, it will feel like a waste", I said to myself. There aren't that many women engineers in my city or my province already, so I would not only be quitting my ideal profession but I would also be leaving my fellow female engineers without a mentor or a guide.

So, I stayed and I grabbed at my engineering career as if it was a lifeboat. It was actually scary to realize that I could have left the engineering field, like many other women do, when faced with a difficult career that has the added challenge of the gender gap. So, I stayed. I stayed, but instead of going back to being an engineer in industry, I switched to being an engineer in academia. I decided to teach engineering because I love it. My best memories came from being in school, from kindergarten all the way to graduate school. So, the thought of getting paid to go back to school was extremely appealing. The question now was: where was I going to teach?

19.8 GRABBING ONTO ENGINEERING WITH BOTH HANDS

Through my networking – that is, contacting my friends and colleagues and telling everyone I was fully available to take on teaching jobs – I have built a very successful new career. Now I teach in various universities as a sessional instructor in engineering and project management, and in two languages: English and Spanish. I can truly say this is my dream job: guiding and supporting my team of students, hundreds if not thousands of students. It turns out that with 20 years of engineering industry experience under my belt, I have become the perfect teacher. My students get the "real" stories about being in the industry, and about thriving as a woman in engineering.

But being a successful engineer did not erase the memories of how close I came to quitting engineering. I just knew I had to share whatever lessons I had learned about supporting EDI in STEM, and this "giving back" became such a strong calling that I felt I had to do something about it. One of the reasons I decided to stay in engineering after leaving my industry job was to become a mentor, so I did the ultimate mentoring job: I opened a non-profit company that organizes conferences to women, and their supporters, in engineering. The mission of the organization was so clear to me: give women, and their supporters, practical tools and advice to help them stay in engineering until retirement. The idea worked, and I was able to gather many supporters. Now, after five years of hosting the women in engineering summit and winning two EDI awards in the process, I am pleased that the mission continues.

The other thing I did to "give back" to my profession was to run for council election for our local engineering association. I had wanted to run for council, which is like the board of governors, since I was in my first year of engineering in Canada. Back then I was told I needed to graduate first and then get at least 10 years of work experience in engineering. The wait was long, but eventually I was able to run, be elected and volunteer as a councillor. The work of the council is to steer the association and protect the title of engineer. I love to volunteer in the council for the association, and I love to uphold the mission of the women in engineering summit.

My three kids are now almost grown – the youngest is already 14 years old – and I am still in love with my husband, plus I make my living as an engineering and project management professor at various universities. The transition from industry to academia has been fun, and I am glad to say I am fully active in engineering-related pursuits. Through all the efforts, sacrifices and challenges, I have been an active engineer for 20 years and I still have another 20–30 years to go. It has been a journey, and what has kept me on the right path has definitely been to "remember the why" I want to be an engineer. I wholeheartedly believe for the world to be successful, it needs more women in engineering and in positions of power, so I wish the readers great success in whatever venture they undertake!

ABOUT THE AUTHOR

Claudia Gomez-Villeneuve, P.Eng, M.Eng, PMP, DTM, FEC, FGC (Hon.), was born in Barranquilla, Colombia. In Canada she was a pipeline project engineer for almost 16 years and now teaches engineering and project management at various universities in both English and Spanish. Claudia won the 2019 APEGA Summit Award as a Champion for Women in Engineering. She lives in Edmonton, Alberta, Canada, with her husband and their three children. You can catch her next hybrid delivery of the Women in Engineering Summit, WES, by visiting www.womeninengg.ca.

20 A Woman in a Man's World or a Person in a Person's World?

Angelica González
Hexpol Compounding, Mexico/USA

20.1 GROWING UP IN A TRADITIONAL MEXICAN FAMILY: ADVANTAGE FOR SOME, DISADVANTAGE FOR OTHERS

Growing up in a traditional Mexican family is one of the most beautiful and nurturing things that could have happened to me. You grow by default with endless values, empathy, humility, respect, hard work, and many other things that help you to face the real world. You don't acquire these elsewhere, certainly not at such a young age; however, it is not easy to break or run away from stereotypes. In my case, I had to go from growing up in a family where studying was not the norm, and even less for a woman. Getting married and being a housewife was the natural path, and sometimes it seems like the only option. Having studied engineering with only men and nowadays living alone, away from my family and traditions, and not even having someone near to turn to while practicing my career in the United States has been a deviation from my traditional Mexican upbringing. However, there is a point when you decide what to do with your "weaknesses"; turn them into strengths, or continue believing that they are ties that hold you back.

I'm 28 years old, I was born and raised in a town called Venadero in Aguascalientes, Mexico. Fortunately, I grew up in a traditional Mexican family, I'm the youngest of four siblings (three women and a man), and the conventional thing in my family is that the man is the one who goes out to work, and the woman stays home and takes care of the family. To my great fortune, I was the youngest and with whom they were not as strict as they were with the oldest.

Studying a career has never been important in my family, but working hard and in the smartest way possible to be able to live well and enjoy life without economic worries was very important. In the same way, they taught me to always be grateful, to do things right, and to treat any task with respect, because every job is important no matter how simple or insignificant it might seem to others, and every step of the way my family was there showing me their support. Since I was little, I was very interested in continuously learning new things, and when I already learnt something, I always wanted to go a little further. When I finished high school, I wanted to go to the city to study because I knew I could find more opportunities there.

DOI: 10.1201/9781032679518-23

At that time, to apply for high school, you had to wait in line and get a card to apply for the exam. People would go a day early to get a card for the technical career they liked the most. My dad didn't want to let me go to the city to study and said that I should go to school in town, that it didn't matter where I studied if I was smart because a kid who wants to learn, will learn anywhere. I think he was right, but sometimes you need to get a little closer to where the opportunities are to continue growing.

A week before the date to get the application form to have access to the admission test, I just kept on insisting to my dad in one and 1000 ways to let me go. It was the night before, when there were already a lot of people lined up, that my dad told me "Go, don't mess around anymore". At 7 a.m. I took the public transportation from my town to the city (for the first time alone) and set out to the train. I got the application for the technical career of "Bilingual Executive Assistant" which was the only one available, I took the exam and was accepted (they had a high rejection rate), and that's when everything began to change.

My high school years were very good in opening my panorama to a different world, to realize that there was more to learn and that there was a world to discover. I entered several contests, and one that made the difference in everything was when I competed to be able to travel to another state to a National of Art and Culture in the discipline of Declamation. My biggest "problem" was getting my dad's permission to go to the other state if I won the state contest.

One night while my parents were watching TV, I approached them and told them about the contest and that if I beat the other participants from the other schools in the state, I would be representing Aguascalientes in Cancun, Quintana Roo. They didn't pay much attention to what I was saying, and I remember my dad's words, "Win the contest first". Those words were like a challenge for me, and I knew that if I won, he could no longer say no because I would go with all expenses paid. I won the contest, and my dad had to let me go alone when I was just 15 years old for the first time to another state, which was not normal or easy for him.

During my high school and college years, one of the ways to earn money was to teach summer courses in my town for regularization in basic mathematics, algebra, calculus, geometry, etc. So, over time, people started to seek me out for advice. I also organized some graduation and social events which were very profitable for me.

20.2   SUPPORT AND ACCEPTANCE: DURING CHANGE AND INTEGRATION, YOU WILL ALWAYS NEED PEOPLE TO BE WITH YOU ALONG THE WAY

On the way you will always find people who do not want the world to change; they just like the way things are because in their heads that is how things are supposed to be. They will put obstacles for you to fall and ridicule you to make you give up. However, you will also find people willing to help you with your day-to-day. People who see everything from a different perspective and who will join you to achieve everything you propose. People who let you follow your path, and if they watch you fall, they offer you their hand to get back on your feet.

It was time to ask for permission to go to college and choose a career. My GPA was very good right out of high school, and there were several options to study at different universities. One of the main goals for me was to get a scholarship and thus make it easier to get permission to study at the university, no matter what the career, although I was very good at mathematics and with a very agile mind to solve problems. The first thing I thought of was to study to become a lawyer, a career where there were many women, but in the end, they offered me a scholarship and a "free pass" if I entered the career of mechatronics engineering because they had a woman integration program and my grades were high. The Universidad Politecnica de Aguascalientes was doing great!

On the day of the interview, the teacher who received me for the interview asked, "Why do you want to enter mechatronics engineering?". My answer was simple and direct, "There is much to learn and I'm interested in being part of the integration of women in this field", to which he replied, "Are you sure you do not want to enter just to find a boyfriend?" I only confirmed that this was not the reason why I wanted to enter the career, but internally it became a challenge for me to finish the career and to do it in the best way possible.

Time went by and I was very frustrated because I did not understand the basic concepts of electricity, control, and mechanics, among other things. A professor encouraged me and told me "You may seem the weakest for now, but with time and the other skills you have, you will stand out". With those words in mind, I dedicated myself to studying outside of my class schedule so that I could keep up with my classmates and not hold them back in learning. When we first started, there were 8 women in the class of about 120 students, and we finished with only 6.

In my second school year, I failed the subject of derivatives and integrals, and I had to take it again to be able to continue with my studies. Obviously, I did not comment on this at home and said only that I had more time to study, because I did not want there to be any reason for not being able to continue studying. After taking the subject again, it became a strength for me because, in my efforts to stay in the game, I invested a lot of time in studying. Afterwards, I did my school service giving regularization classes in the same subject.

During all the years of study at the university, on one occasion a professor of the subject of stress analysis (who did not like that women were in the class and was very contemptuous of us) he called us by our name, but masculinizing them, did not accept our opinions and made offensive comments to us. We felt what was happening was really offensive, but we did not want to say anything; we were finally in the "man's world".

A very good friend of mine, after the professor openly made a vulgar comparison about me in front of everyone in the class, raised his voice and said that this was not right; several classmates agreed with him and denounced him to the directors for us. The university, after investigating, terminated the professor's contract. At that moment I didn't realize what was really happening, I was so caught up in working to reach my dreams that I let others speak up for me. Don't get me wrong – it was a great show of support from my friend and our classmates not to allow that situation to happen.

20.3   ACT STRATEGICALLY, MAKE EVERY EFFORT WORTHWHILE

As time goes by, you have to identify your strengths and weaknesses; you have to detect what works for you so you can act strategically with a plan in mind. You have to know what you need to get to where you want to be, to be what you want to be.

While all this was going on at the university, it was becoming more and more difficult to be able to do the team projects. This meant staying up late, and that was something that in my house, they did not feel comfortable with because I was studying with only men. One way or another we divided the work and made it work. A couple of times my coworkers had to go to my house (far from the city) to work there to finish the projects.

Time went by, and my strategy of joining different competitions to be able to travel to other states and learn as much as possible was still strong at the university. I was part of the American Society of Mechanical Engineers (ASME) and participated along with other classmates in the "Human Powered Vehicle Challenge" contest, where the main objective was to apply engineering and design a human-powered vehicle that would be competitive. It was like being in Formula 1 of pure human-powered cars!

An internal competition was held in order to represent the university in the National Competition of Sumo Robots and Line Followers in Queretaro, Mexico. My team and I won and had to prepare ourselves to compete with the best engineering schools in the country. Unfortunately, we didn't get any place, but it was definitely a very comprehensive learning experience.

A year before finishing my degree, I enrolled in the Math Olympiad in the individual category, which consisted of two stages. The first was a general exam in front of a computer with supervision and the final stage with the three best scores. I qualified for the final, and this stage was live in front of the university students, with one camera focused on me and the other on my scores. We had a button to press once we had the answer to the question. I won first place.

During my college years, I was involved in many other things that were not directly related to my career but that gave me access to people with extensive knowledge and a lot of experience. I participated in several university events giving a thank you speech on behalf of the student community to companies that donated machines and robots so that students had the opportunity to learn by practicing. Likewise, along with other colleagues, we led two annual events for the Day of the Mechatronics Engineer where we coordinated all the conferences, workshops, and conviviality for three days for all levels of the career.

20.4   PERSEVERANCE AND CUNNING: THE BEST ALLIES IN ANY SITUATION

In Mexico, when you finish all your curriculum, the next step is to do your internship, where you have to do a project and your thesis to get your degree. The government of the state of Aguascalientes presented a call for scholarships to study abroad, the requirements were to have a certain GPA, the ability to communicate in the language of the country you were going to, and a letter of acceptance from the university of that country.

I decided to start the process without telling my family because I knew it was something they would not accept easily. I was already going to finish all my courses, and the challenge here was to be able to do my internship in another country but to be accredited in order to obtain a scholarship as a student. I contacted two doctors from the University of Puerto Rico at Mayaguez to request to do my internship with them, doing research on a project in which the main objective was the development of a portable high-temperature oven that would allow students to practice and manufacture their parts from a friendly interface using Labview and PLCs. After insisting and finding a way to make this a win-win, I got the approval to obtain credits for doing the research and development of the project and also to be able to present it in my thesis for my degree.

A few weeks before the scholarship results were announced, I told my family that I had applied for the scholarship and that they were going to pay almost all my expenses during my stay in Puerto Rico. They took it well at first, I think because they thought they would not give me the scholarship. The day of the results we were watching TV and together with other classmates we were watching the publication. When I saw my folio number in the scholarship winners I felt joy, but also a lot of fear, because it would be something totally new.

My family, although more accessible and trying to accept the idea of my ideals, insisted that I should not accept the scholarship because it would be six months of wasted time. I decided to take it and go, and my family supported me in spite of everything.

The stay in Puerto Rico was where I got to see the importance of values and integrity of people. It is very easy to lose your temper and let yourself be guided by many mundane things. I enjoyed, studied, cried, and valued many things. I returned home with a much broader vision and more confidence than ever, and I wanted to continue exploring other cultures.

20.5 INDUSTRY: CONTINUOUS CHALLENGES

The industry is an environment where you have the opportunity to learn from many with a lot of experience. You will come across those who want to help, share information without anything in return, but you will come across others who don't want you in that world, who don't think you belong. The important thing here is that you keep your morals and ethics so strong that you have nothing left but to rely on your talent. It is here that I valued growing up in a traditional Mexican family, where I had to fight for what I wanted with values and talent.

It is interesting how time goes by and you find and trace your path. Upon returning from my stay in Puerto Rico and finishing my college thesis, I went to work in the industry where I learned many things that I value today, but I continue to struggle.

I have had to face different situations where, to be honest, I didn't know how to deal with them, and I decided to shut up and work harder. Several suppliers offered to help me grow quickly with their contacts in the industry if I agreed to have "something" with them. There were suppliers who did not want to review any technical details with me because they believed I did not have the capacity or the technical knowledge, or simply because "they did not work with women". It was something that I decided to keep quiet and continue doing my job, letting my work and results speak for me, while continuing to grow in the industry.

My entire career in the industry has been related to excellence in maintenance, engineering, and continuous improvement. I started as a maintenance technician, but with time other skills were seen and I was promoted to Maintenance Planner, helping with spare parts inventory control, planning, scheduling, and maintenance purchasing. Sometime later, still looking for opportunities that would help me grow, I got involved in researching topics of interest to the company where I worked. Eventually, I was promoted to be in charge of a CMMS system in several locations and not only for the site I worked in.

There have been great leaders and friends who have been teaching me, guiding me, and helping me to grow in this industry.

A couple of years ago I had an offer from one of the best and youngest leaders that I ever had to work for, from the same company in the United States as a maintenance systems engineer. I would be working along the different sites in Mexico and the United States with all maintenance managers and maintenance teams helping to drive the company to maintenance excellence and Reliability.

Time goes by, and you will always be in the midst of different situations that you have to face, and fears that grow along the way. The key is to keep on going, be faithful to yourself and your beliefs, allow yourself to just BE, keep trusting people, keep trusting yourself, keep demonstrating that anything is possible, keep learning and investing in your mind.

20.6   MY ADVICE FOR WOMEN IN THE INDUSTRY

During these seven years in the industry, I have been in constant preparation with different courses, finances, leadership, project management, Excel, Power BI, data analysis, and security, among many others. I know that I'm a young person, eager to grow and to continue opening doors and leading the way for women and Mexicans whenever I go, so here are some tips that I heard from someone and I try to follow myself every day:

- *Always think of yourself as part of the situation.* Going into a problem assuming that you must be treated differently because you are a woman will make it very difficult for you to be confident and perform to the best of your ability.
- *Surround yourself with like-minded people.* Seek out proactive, productive, intelligent, and honest people and learn as much as you can from them.
- *Control your mind.* Visualize what you want, learn to discover what you need to achieve it, and execute with a plan in mind.

- *Be true to your beliefs.* Set high standards of ethics, discipline, and passion for what you do, don't let anyone try to undermine you, and celebrate yourself every step of your career.
- *Own the way you act.* Be smart and put your assets first, do things that reflect who you are, be ethical, and be honest with yourself and the world.

It's not about being a woman in a man's world, it's about being a person in a person's world! See you on the road!

ABOUT THE AUTHOR

Angelica González, at 28 years of age, is a first-time author. She grew up in a traditional Mexican family in Aguascalientes, Mexico. She has a bachelor's in mechatronics engineering from the Polytechnic University of Aguascalientes and an MFA in web design and digital marketing from the Cuauhtémoc University of Aguascalientes. After starting in the engineering world, she embarked upon a career as a self-driver, and an active continuous improvement practitioner in the maintenance and reliability industry. Along her way from high school, college, and her personal life, she has been such a big inspiration to women who know her. She is an example of perseverance, consistency, and inclusion for women in a "man's world". She currently resides in Ohio, USA, where she works in the rubber industry as a maintenance systems engineer working with reliability and maintenance excellence.

21 The Complex Reality

Unraveling the Underrepresentation of Women in STEM

Genevieve Cheung
Atticus Consulting, Canada/Barbados

21.1 EXPLORING GENDER DIVERSITY

Gender diversity in STEM fields is a topic that has gained significant attention in recent years. The underrepresentation of women in science, technology, engineering, and mathematics has sparked numerous discussions and initiatives aimed at understanding the reasons behind this gap and finding ways to bridge it. As a female civil engineer who has spent most of my career in male-dominated industry and workplaces, I feel qualified to explore the various factors that contribute to this issue, and attempt to delve into the complexities surrounding women in STEM.

The presence of women in STEM is crucial for several reasons. Firstly, it promotes diversity of perspectives and ideas, leading to more innovative solutions and advancements in these fields. Secondly, it challenges gender stereotypes and biases, dismantling the notion that certain careers are exclusively suited for men. Lastly, it provides opportunities for women to thrive in high-paying and influential professions, empowering them to contribute to society in meaningful ways.

To understand the gender gap in STEM, it is important to examine the statistics and representation of women in these fields. While progress has been made, women continue to be underrepresented, especially in certain STEM disciplines such as engineering, mathematics, and computer science. This begs the question: why are there fewer women in these fields?

Contrary to popular belief, the explanation does not solely lie in sexism. While it is true that sexism can influence some women to opt out of engineering or mathematics, it does not fully account for the overall gender gap in STEM. There are other plausible explanations that deserve consideration.

One such explanation revolves around the idea that men and women may have different interests and preferences, whether by nature or nurture, leading them to choose different career paths. Research suggests that, on average, women tend to gravitate towards professions that involve working with people, while men are more drawn to careers that involve working with things or systems. This divergence in interests could contribute to the disparity observed in specific STEM fields.

DOI: 10.1201/9781032679518-24

Additionally, achieving gender parity in one field without it being mirrored in another can be statistically improbable. Assuming overall gender parity, it is statistically impossible for women to be the majority in one broad field without simultaneously being the minority in another. Understanding this statistical reality is crucial when considering the representation of women in STEM.

By exploring these alternative explanations and challenging conventional thinking, we can gain a more nuanced understanding of the gender gap in STEM. This chapter aims to foster an open and honest dialogue about the complex nature of this issue and to encourage further exploration of potential solutions. Through a deeper understanding of the barriers faced by women in STEM, we can work towards creating a more inclusive and diverse environment that benefits everyone involved.

While it is important to acknowledge these accomplishments, it is equally crucial to recognize that challenges still exist for women, especially in non-Western countries and in terms of female promotion to senior positions. However, by highlighting the positive progress made in educational attainment, we can build upon these achievements and continue working towards gender equality in all areas of society.

21.2 WHAT IS THE GENDER GAP IN STEM?

The gender gap in STEM fields is a well-documented phenomenon that has garnered significant attention in recent years. Despite efforts to promote diversity and inclusion, women continue to be underrepresented in various STEM disciplines. Statistical data consistently shows that women are underrepresented in fields such as engineering, mathematics, and computer science. While there has been progress in increasing the number of women pursuing STEM careers, the gender gap remains prevalent. This is evident in the lower percentage of women enrolled in STEM programs at universities, as well as the limited representation of women in leadership positions within STEM industries.

Several factors are attributed to the gender gap in STEM. One significant factor is *societal and cultural influences*. From an early age, children are exposed to gender stereotypes that shape their perceptions of what careers are suitable for men and women. Girls are often steered towards more nurturing and social fields, while boys are encouraged to pursue technical and scientific disciplines. These stereotypes and biases can influence career choices and discourage women from pursuing STEM fields.

Another factor is the *lack of female role models and mentors in STEM*. The underrepresentation of women in these fields means that aspiring female scientists and engineers may have limited access to relatable role models who can inspire and guide them. The absence of mentors can make it challenging for women to navigate the male-dominated STEM landscape, hindering their career advancement and sense of belonging.

Furthermore, *workplace culture and biases* play a significant role in perpetuating the gender gap. Studies have shown that women in STEM often face gender bias, stereotyping, and unequal treatment in the workplace. These biases can lead to feelings of isolation, imposter syndrome, and a lack of confidence among women, which may contribute to lower retention rates and slower career progression compared to their male counterparts.

The implications of the gender gap in STEM extend beyond individual careers. A lack of gender diversity in these fields hampers innovation and creativity. Different perspectives and experiences are crucial in problem-solving and developing robust solutions. By excluding a significant portion of the population from STEM fields, we limit the potential for groundbreaking discoveries and advancements.

But are these really the root causes of gender gaps in STEM? Let's start over from the well-accepted, undisputed statement earlier in this section that women are underrepresented in STEM disciplines.

21.2.1 Determining Some Root Causes

In the United States, women have surpassed men in earning PhDs over the past decade, demonstrating a significant achievement and challenging long-held sexist stereotypes. Figure 21.1 shows recent data highlighted in an article published by the World Economic Forum that indicates that more women than ever are making original contributions to knowledge through their doctoral research, edging out men by more than 3 percent as a majority. On the surface, this is undoubtedly a positive trend and a testament to the capabilities of women in academia.

However, a closer analysis of the figures reveals that men continue to dominate in specific fields such as engineering (76.6 percent), mathematics and computer sciences (74.9 percent), and physical and earth sciences (65.9 percent). These statistics emphasize the need for continued efforts to encourage women to pursue STEM subjects at higher levels, but doesn't seem to offer any new insight to the pervasive narrative that women are underrepresented in STEM fields.

U.S. Women Earned More PhDs Than Men Last Year

Doctoral degrees awarded by field and gender (2016–2017)*

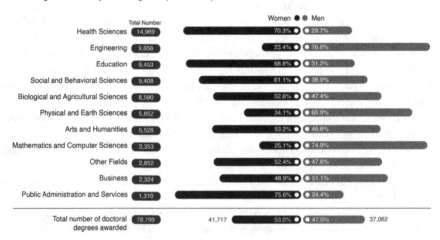

*Because not all institutions responded to all items, details may not sum to totals

FIGURE 21.1 Doctoral degrees awarded by broad field and gender in the United States, 2016–2017.

Firstly, it is important to celebrate the significant achievement of women in attaining higher levels of education, including earning more PhDs than men. This narrative should be recognized and acknowledged as a testament to the capabilities and achievements of women in academia. However, it is also crucial to address the need for a balanced approach in encouraging both men and women to pursue fields of study that are traditionally dominated by the opposite gender. Just as efforts are made to encourage women to study STEM subjects, there should also be a focus on encouraging men to pursue health or social sciences at a higher level. By promoting gender diversity in all fields of study, we can create a more inclusive and equitable educational landscape.

Secondly, the definition of STEM needs to be more clearly defined and inclusive. The commonly understood perception of STEM tends to focus on "hard" sciences like physics and chemistry, as well as technology-related fields like computer science and engineering. However, this narrow definition overlooks the contributions and importance of other fields such as health sciences, social and behavioral sciences, and biological and agricultural sciences. These disciplines also involve scientific research and utilize technology in their respective domains.

How is a dentist's drill not technology? How is an X-ray machine used in radiography or the heart and breathing monitors used by nurses in an intensive care unit not technology? How is a whiff of oxytoxin used in a psychology experiment not technology? How is examining and modifying DNA in biology and agricultural science not technology? By adopting a more accurate and inclusive definition of science and technology, we can recognize the valuable contributions of women in these fields and eliminate the perception of underrepresentation.

21.2.2 REDEFINING THE STEM DEFINITION

When a broader definition of STEM is applied, considering the inclusion of health sciences, social and behavioral sciences, and biological and agricultural sciences – referring the same data in Figure 21.1 data reveals that women earn 50 percent of PhDs in STEM fields. This balanced representation across different disciplines demonstrates that the goal of achieving gender parity in STEM can be accomplished when STEM is accurately defined.

You would notice that I have introduced an additional classification. I have made a broad judgment that engineering, physical and earth sciences, and mathematics and computer sciences, the narrower definition of STEM, are primarily focused on *things*. On the other hand, health sciences and social and behavioral sciences, the sciences often overlooked when discussing STEM, are more focused on *people*. Biological and agricultural sciences, which I consider to be more about animals and plants, serve as a "halfway house" between the inanimate things studied in the "hard" sciences and the people studied in the health sciences and social and behavioral sciences.

As the data shows in Figure 21.1, this "halfway house" field has the most balanced participation rates between males and females. This seems to confirm that males and females have roughly equal interests in studying animals and plants. Males as a group are less inclined towards people and more towards things, while females as a group tend to be more interested in people and less interested in things. This finding aligns with well-known psychological research on the people/things asymmetry between the sexes.

However, this does not mean that a particular woman is incapable of studying things, or that women cannot code, for example. Nor does it imply that a particular individual, regardless of gender, cannot be interested in both people and things. The preference for *people* or *things* exists on a spectrum and is not a simplistic binary. The figures simply demonstrate that, as a group, some women tend to be somewhat more interested in the so-called "soft" sciences and somewhat less interested in the so-called "hard" sciences. Furthermore, the distinction between hard and soft sciences is outdated and oversimplified. Therefore, there is no significant problem here, just a reaffirmation of a well-established psychological finding regarding people and things.

21.2.3 THE SEE-SAW PRINCIPLE

This ties back to my earlier point about not emphasizing the need to increase the representation of men in female-dominated fields. If we assume overall parity between the numbers of men and women and that approximately equal numbers are available to enter any broad field of PhD study, it becomes mathematically impossible to have majorities of women in some sectors (e.g., health sciences and education) without having minorities of women in other sectors (e.g., engineering and computing). Let's call this the see-saw principle.

To provide a visual representation, imagine we allow 100 kids into a park, with 53 girls and 47 boys. In the park, there are 11 see-saws. If the girls rush towards the 4 see-saws they prefer for various reasons, and 4 see-saws have roughly equal numbers of boys and girls, it should not be a cause for concern when we come across 3 see-saws with relatively few girls on them. This simply means that there are not enough girls left to occupy those see-saws. The majority of girls freely chose to play on the 8 other see-saws based on their preferences and interests.

We should not argue that there is something malevolent about the 3 see-saws with mostly boys compared to the 4 see-saws with mostly girls, especially when we have data showing asymmetries in preferences for certain types of see-saws between the sexes. Such data exists. If we extend this analogy to academia, we can see that see-saws involving people are more appealing to females, while those involving things are more attractive to males. Some see-saws, such as biological and agricultural sciences, business, arts and humanities, and other fields, are equally attractive to both males and females.

Based on the analysis I have presented thus far, we can conclude that sexism is not a particularly convincing explanation for the relative lack of women in the narrow definition of STEM at universities. I want to emphasize that I do not doubt that sexism may explain why some women opt out of engineering or mathematics. However, it is plausible to consider that the relatively low numbers of women in the narrow definition of STEM fields could be attributed to a general preference among females for working with people rather than things. Additionally, the statistical impossibility of women being a majority in one broad field without being a minority in another field, assuming overall parity, also contributes to this pattern.

It is essential to challenge the notion that sexism is solely responsible for the gender disparity in STEM. While sexism may play a role in certain aspects, it does not account for the entire picture. By acknowledging and understanding the different preferences and interests between males and females when it comes to academic fields, we can have a more nuanced perspective on the representation of women in STEM.

Rather than focusing solely on increasing the representation of women in traditionally male-dominated fields, we should strive for a balanced approach that encourages individuals to pursue their own interests and passions, regardless of gender. This means recognizing and valuing the contributions of both men and women in all fields of study, whether they are predominantly focused on people or things.

By broadening the definition of STEM to include disciplines such as health sciences, social and behavioral sciences, and biological and agricultural sciences, we can create a more inclusive framework that acknowledges the diverse range of scientific endeavors. This expanded definition reflects the reality that these fields involve scientific inquiry, research, and the application of technology. It also aligns with the balanced participation rates observed between males and females in these fields.

21.3   THE EQUALITY AGENDA VERSUS THE EQUITY AGENDA

So how does this translate into the workplace? Based on the aforementioned description, it appears that we have not a gender representation issue in STEM as a whole, but rather a lower representation of women within the narrow definition of STEM, particularly in fields such as engineering, mathematics, and computer sciences. However, the concept of diversity and inclusion is still widely debated in our society, and there are differing perspectives on what it truly means.

The conversation surrounding gender diversity in STEM fields often involves two distinct approaches: the equality agenda and the equity agenda. The *equality agenda* advocates for equal opportunities and treatment for individuals regardless of their gender, race, or other immutable characteristics. It promotes the idea that the most talented and passionate individuals should have access to and succeed in their desired STEM fields, irrespective of their life circumstances or identities. Diversity, within the framework of the equality agenda, encompasses a broad range of perspectives and experiences, including racial, ethnic, and gender diversity, as well as diversity of political and religious beliefs.

Inclusion, as *defined by the equality agenda*, involves welcoming and encouraging individuals from underrepresented groups to pursue male-dominated STEM fields as careers. It focuses on removing artificial barriers that may prevent individuals from accessing and excelling in these fields. The equality agenda aims to create an environment where individuals are evaluated solely on their merit, abilities, and dedication, rather than being limited by societal biases or discriminatory practices.

On the other hand, the *equity agenda* goes beyond the pursuit of equal opportunities and challenges existing power structures. It seeks to redress privilege and correct historical and systemic inequalities by advocating for unequal outcomes as evidence of the need for change. In this context, diversity is not only about representation but also about dismantling structures that perpetuate inequality. *Inclusion*, within the *equity agenda*, involves creating a culture that promotes and prioritizes historically marginalized voices, even if it requires the exclusion of certain ideas and opinions.

The equity agenda recognizes that achieving equal representation in STEM fields may not be possible without addressing underlying power imbalances and structural barriers. It acknowledges the need for deliberate efforts to level the playing field and provide additional support and resources to underrepresented groups. Advocates of the equity agenda argue that until perfect gender parity is achieved, diversity programs and initiatives should be implemented to address the inequities faced by women in STEM.

The debate between the equality and equity agendas raises important questions about the most effective strategies for promoting gender diversity in STEM. While the equality agenda emphasizes equal opportunities and meritocracy, the equity agenda emphasizes the need to address systemic barriers and power imbalances. Both agendas have their merits and challenges, and finding a balance between the two is crucial for creating inclusive STEM environments.

21.4   IT'S COMPLICATED

To truly bring about meaningful change, we must engage in a deeper discussion about the barriers that contribute to the underrepresentation of women and minorities in STEM. The issue at hand is complex, and while we have established that gender does not determine capabilities, there is ample research to support the notion that men and women often have different interests that lead them to choose different career paths. This section explores some of the key factors that contribute to the underrepresentation of women in STEM fields, including societal expectations, early childhood experiences, confidence and imposter syndrome, and work-life balance.

   Societal Expectations: From an early age, girls are often socialized to believe
      that certain careers or fields are more suitable for boys, while others are
      more appropriate for girls. This societal expectation shapes their interests,
      choices, and career aspirations. The perception that STEM fields are pre-
      dominantly male-oriented can discourage girls from pursuing careers in
      these areas.
   Early Childhood Experiences: Early childhood experiences play a crucial role
      in shaping individuals' interests and career paths. Research suggests that
      boys are more likely to be encouraged to explore and engage with STEM-
      related activities, while girls are often directed towards activities that focus
      on social interaction and caregiving. These early biases can influence the
      development of skills and interests that align with STEM fields.
   Confidence and Imposter Syndrome: Women often face challenges related to
      confidence and imposter syndrome in male-dominated fields like STEM.
      Stereotype threat, a phenomenon where individuals feel at risk of confirm-
      ing negative stereotypes about their group, can affect women's confidence
      in their abilities. *Imposter syndrome*, the feeling of being a fraud despite
      evidence of competence, disproportionately affects women and can hinder
      their pursuit of STEM careers.

Work-Life Balance: Balancing personal and professional responsibilities is a concern for individuals in various fields, but it can have a particular impact on women in STEM. The demanding nature of STEM careers, long work hours, and limited flexibility can pose challenges for women who prioritize family and work-life balance. These factors may influence career choices and contribute to the underrepresentation of women in certain STEM disciplines.

While it is true that men have had the advantage of shaping a culture that aligns with their values and interests due to their earlier participation in STEM fields and the workforce in general, it is unfair to place the blame solely on men and the patriarchal organizations they have built. While there are certainly instances of problematic cultures and bad actors, the majority of women work in companies that make significant efforts to provide a supportive work experience.

It is essential to recognize that the gender gap in STEM is not solely a result of individual preferences or choices. It is a complex interplay of societal norms, early experiences, confidence issues, and work-life balance considerations. Understanding these factors can help inform strategies to promote gender diversity in STEM fields.

21.5   OVERCOMING BARRIERS AND SHIFTING THE NARRATIVE

In order to address the underrepresentation of women in STEM, it is crucial to acknowledge and overcome the barriers that hinder their participation. At the same time, it is equally important to shift the narrative surrounding women in STEM and celebrate their achievements and contributions. By combining these efforts, we can create a more inclusive and empowering environment for women in the field.

One of the key barriers that women face in STEM is the *lack of encouragement and support from an early age*. Societal norms and stereotypes often influence the choices and aspirations of young girls, steering them away from STEM subjects. This can result in a significant gender gap in educational pathways and career choices later in life. To overcome this barrier, it is essential to provide girls with equal opportunities and exposure to STEM from an early age. This can be achieved through initiatives such as mentorship programs, outreach activities, and educational campaigns that challenge gender stereotypes and inspire young girls to pursue their interests in STEM.

Another significant barrier is the *persistent gender bias and discrimination* that women encounter throughout their STEM journey. This can manifest in various ways, including unequal treatment, unconscious bias, and a lack of opportunities for career advancement. It is crucial for institutions and organizations to actively address and eliminate these biases through policies, training programs, and inclusive hiring practices. By creating a supportive and inclusive culture, women in STEM can thrive and contribute their unique perspectives and talents to the field.

Furthermore, the issue of *work-life balance* often poses challenges for women in STEM. Balancing family responsibilities and career aspirations can be particularly demanding, leading some women to opt for career paths with greater flexibility or lower time commitments. It is important for organizations to provide supportive policies and practices that promote work-life balance, such as flexible work arrangements, parental leave, and childcare support. By creating an environment that values and accommodates the diverse needs of women in STEM, we can retain their talent and expertise in the field.

Shifting the narrative is equally crucial in celebrating women in STEM and inspiring future generations. It is essential to showcase the achievements and contributions of women in the field, highlighting their groundbreaking research, innovative discoveries, and influential leadership roles. By amplifying their stories and voices, we can challenge the stereotype that STEM is a male-dominated domain and inspire young girls and women to pursue their passions in these fields. Recognizing and celebrating the accomplishments of women in STEM also helps to dismantle the notion that women are inherently less capable or interested in these disciplines.

21.6 THE COMPLEX, MULTIFACETED CONVERSATION ABOUT GENDER DIVERSITY IN STEM FIELDS

The conversation about gender diversity in STEM fields is complex and multifaceted. While it may be difficult to achieve complete gender parity in engineering, mathematics, and computer sciences, it is crucial for our community to be honest and realistic about the challenges we face. Women have the capability to build things and excel in coding, but it is also true that not all women may have a strong interest in these fields, leading to a potential imbalance.

However, accepting this reality does not mean that women should feel unwelcome or undervalued in STEM. On the contrary, recognizing that women may be in the minority should encourage organizations to appreciate and celebrate the women who do choose to pursue careers in these fields. It is essential to focus on retaining and supporting these talented individuals so they can succeed in their chosen paths.

Additionally, efforts to promote gender diversity should not be limited to attracting more women to STEM fields. It is equally important to create an inclusive environment that supports the retention and advancement of women in STEM careers. This involves addressing issues such as unconscious bias, providing mentorship and support networks, and promoting work-life balance.

Throughout this chapter, we have explored various factors that contribute to the underrepresentation of women in STEM, including societal stereotypes, lack of support and mentorship, and work-life balance challenges. By addressing these barriers and promoting inclusivity, we can create an environment where everyone can thrive and contribute their unique perspectives and talents.

In conclusion, while achieving complete gender parity may be challenging, our focus should be on creating an inclusive and supportive culture where women in engineering, mathematics, and computer sciences are valued and given equal opportunities to succeed. By embracing diversity and fostering an inclusive environment, we can harness the full potential of our STEM workforce and drive innovation and progress.

ABOUT THE AUTHOR

Genevieve Cheung, P.Eng, MBA, is an accomplished project manager known for her expertise in the construction industry. She holds a bachelor's degree in civil engineering from the University of Toronto and a master's degree in business administration from IE Business School. Genevieve's professional journey began as a consulting engineer in Canada, where she gained valuable experience and obtained her Professional Engineer (P.Eng) license. She moved to the Caribbean about 10 years ago for a work opportunity, and specialized in project management for hospitality construction projects. Her niche lies in providing procurement management services, ensuring seamless project delivery from pre-construction to completion. As a woman in a predominantly male-dominated industry, Genevieve has firsthand experience navigating the challenges and stereotypes present in the construction world. She is on a mission to challenge the notion that construction is solely a male domain. Genevieve firmly believes that promoting gender diversity in this industry is not just about checking boxes for diversity's sake but recognizing the unique skills and perspectives that women bring to large and complex projects. With her extensive experience and passion for empowering women in STEM, Genevieve is a trailblazer and advocate for breaking barriers and shattering glass ceilings in the construction field. Genevieve started her consultancy Atticus Consulting in 2022, providing project management and procurement management services to contractors and owners.

22 Why Is Empowering Women in STEM So Important?

Joel Leonard
MakesboroUSA, USA

The world is in the midst of major generational and technological transitions. To avoid preventable disasters and production losses, the world needs to transcend traditional norms by breaking stigmas and stereotypes that prevent the entry of women into the skill tech and engineering sectors. There are millions of unfilled jobs that could easily be filled with capable, hard-working women if properly guided and developed.

22.1 TAKE THIS JOB AND LOVE IT!

Have you ever been to a conference that changed your life? A conference that inspired you not to just go through your normal routine but actually want to do more about a problem that is plaguing society and do things you never would have ever imagined doing? A problem that inspired you to learn all kinds of methods to help get the word out so more can solve the problem? Well, this is what happened to me, Joel Leonard, aka The Maintenance Evangelist, The Makers Maker.

Over 20 years ago, in October 2002, in Nashville, Tennessee, to support the maintenance management training company, I attended the Society for Manufacturing Reliability (SMRP) Annual Conference. In the conference hotel over 600 top engineering, reliability, and maintenance operation leaders attended to discuss the issues, challenges and technologies, processes and systems to manage large, complex logistics businesses. Attendees included top management from Michelin, USPS (the nation's largest fleet at the time and high-speed mail processing centers), Conagra, DuPont, PPG, Conoco, and numerous other manufacturers and power companies.

To kick off the 2002 conference, the now-deceased editor of *Maintenance Technology* magazine provided a stimulating keynote and asked "How many of you are going to retire in the next 10 years?" That question made everyone take note.

When you see some of the best of the best, over 90 percent of the audience, raise their hands stating they would be departing soon, this was very impactful. What's amazing is how many of these major corporations did not have a formal succession plan: a large pipeline of talent that was being groomed to take on the responsibilities because of cost cuts. "Lean and Mean" was the slogan then. Attendees discussed the challenges at getting young talent to take on more responsibility and the negative

DOI: 10.1201/9781032679518-25

image of the profession. The term *maintenance* was a turn-off, associated with just fixing toilets and not as prestigious as being a lawyer or a doctor. Shop was taken out of schools, and everyone was encouraged to pursue four-year degrees regardless if there was a job when complete. Parents frowned upon sending students to trade schools. Companies were shipping US manufacturing to cheap-labor, minimal-government-regulation countries. Everyone was fearful of their future and would not encourage their own children to consider manufacturing in the United States.

Maintenance management was not even deemed by society as a profession. So all of these highly paid leaders had chips on their shoulders, as they did not feel appreciated or valued for all of their hard work and contributions they make to keep the machines working, businesses safe, product commitment fulfilled, products made all require proper maintenance (to fix failures and to prevent failure is called *reliability*).

After sitting in a conference for six straight hours, I decided to see some of downtown Nashville. I walked around the corner of the hotel and saw over 5000 teenagers meandering in lines outside the convention center bundled up to protect themselves from freezing temperatures, singing various songs. At the same time, over 600 top engineering leaders were wondering where their future leaders would come from. I saw 5000 kids thinking that their future is singing and entertaining, and all wanted a shot at being the next American Idol.

That evening, I met with some of the other conference attendees and got some barley-infused beverages. We started talking about the keynote address and what could be done to get more people to pursue these vital jobs. Just telling engineers that we need more engineers is not going to get us enough engineers. So what do we do? Someone said write a book. Good idea, but no one reads books anymore. Someone said publish columns in magazines. Good idea, but that still won't help as much as needed. Then I suggested something that was deemed at first to be absolutely crazy: Why not write a song? Those kids were out there in freezing temps singing songs; imagine if they heard about the issues and challenges about these jobs via music? If a song helped more people think about it, then perhaps some will consider these pathways after all. After my idea was dismissed, a friend said, "You know, Joel has a good idea, but he will never do it."

Those were the magic words. Despite not ever having written a song before, I was determined to prove him wrong. When I got home, I started putting words together, trying to make it rhyme but also put a message that would stick and make people think about what is going to happen in a world without proper maintenance. HOW SAFE DOES IT MAKE YOU FEEL?

Then, the Maintenance Crisis Song was conceived and has been recorded in over 15 genres of the same message.

Years later, while speaking before middle school, I played the follow-up song, "Find Me a Maintenance Woman," to bust gender biases. To my surprise, three of the students already knew the lyrics because a Raleigh country music radio station played it regularly. Then, I was able to go into details of why maintenance function is so important and offered potentially very lucrative options.

I later discovered that when Lockheed Martin hired the first female maintenance manager, leadership played "Find Me a Maintenance Woman" at a company announcement ceremony. So these songs have made an impact in getting the message in the minds and hearts of people around the world. Hope you enjoy and share, so more will help us fix it forward!

Maintenance Crisis Song

No one wants to work in the boiler rooms
No one want to work with the tools
Nation's youth are taking the easy way out
There's no one left to fix our schools
Maintenance technicians are about to retire
Company executives have got no one to hire
How safe does it make you feel?
How safe does it make you feel?
America needs no more TV idols
The talent shows are rigged anyways
Our building maintenance infrastructure could blow or leak or fade anyday
There's no easy road to getting rich, while our national treasures are going
     down in a ditch
How safe does it make you feel?
How safe does it make you feel?
Working in maintenance is an honorable field
Gone are the days of Bubba, Skeeter and Neil
Todays you use computers and sophisticated tools
But there's nobody entering our technical schools
How safe does it make you feel?
How safe does it make you feel?
Supply fans and cooling towers are breaking down all over the place
High pressures, low pressures, springing a leak,
Well, I can tell you it's a big disgrace!
Industry know what we got to do
Find and train a workforce who will fix it for you
How safe does it make you feel?
How safe does it make you feel?
How safe does it make you feel?
How safe does it make you feel?
What if we train them and they leave?
What if you don't train them and they stay?

"Maintenance Crisis Song," Lyrics Copyright 2003
Lyrics written by **Joel Leonard**

This song has been recorded now in over 12 genres, including garage band, blues, hip-hop, gospel, opera, reggae, bluegrass, rock 'n' roll, funk, Greek, and more

Find Me a Maintenance Woman

You may laugh!
You may jeer!

But you can keep your Britney Spears!
I need a woman that can work with gears!
I need to find me a maintenance woman!
I can't use a girl like J Lo!
See I need a woman that can make things go!
One that knows how dc current flows!
I need to find me a maintenance woman!
No supermodels or beauty queens,
Or some American idol who thinks she can sing
I know there are few and far between that can fix my broken hydraulic machines
There's only thing for businesses to do is to have a few women to pull us
    through
So hey there Mr Bossman, won't you find me a Maintenance Woman!
I don't need no girly from Hooters, I need a super duper troubleshooter
Maybe she can fix my desktop computer,
I need me a maintenance woman
She don't need to work with Donald Trump
She can help us get out of this slump
Give our electric generators a jump
We need us a maintenance woman
When we add some women to our staff
You know those other fellas won't catch no flap
They won't need have to open their traps
She ain't going to take any crap
The best thing a good business could do, is to hire some women and that could
    pull us through
Hey there Mr Bossman won't you find us a Maintenance Woman
I said hey hey there Mr Bossman find us a Maintenance Woman
I said hey hey there Mr Bossman find us a Maintenance Woman
I said hey hey there Mr Bossman find us a Maintenance Woman
I said hey hey there Mr Bossman find us a Maintenance Woman
I said hey hey there Mr Bossman find us a Maintenance Woman

"Find Me a Maintenance Woman" – A SkillTV.net Music Video
Final Version Lyrics written by **Joel Leonard**

Take This Job and Love It!

Take This Job and Love It! Soon boomers won't be here anymore!
If you put your heart in it, your job 'comes more than just a chore
Want to fix your financial mess, just clean up your act and go pass a drug test
Soon boomers won't be here anymore. Take this Job and love it.
If you put pride in your work, your job will become something you will covet
Companies need new workers to grow up and show up
So take this Job and love it, Soon boomers won't be here any more
Want to fix your financial mess, clean up your act and pass a drug test. So
    don't ghost and give this job your most!
Take This Job and Love It! Soon boomers won't be here anymore!

If you put your heart in it, your job 'comes more than just a chore
Want to fix your financial mess, just clean up your act and go pass a drug test
Soon boomers won't be here anymore. Take this Job and love it.
If you put pride in your work, your job will become something you will covet
Companies need new workers to grow up and show up
So take this Job and love it, Soon boomers won't be here any more
Want to fix your financial mess, clean up your act and pass a drug test. So
    don't ghost and give this job your most!
Take this job and love it! Take this job and love it! Take this job and love iiiiit!!!

Lyrics Copyright 2022 Written by **Joel Leonard**

22.2   MENTORING WOMEN THROUGH STEM

During the pandemic, we all learned it can be easier to communicate with people from around the world rather than to visit your neighbor across the street by simply pushing a button. So during the pandemic, and even now, I help offer career development advice to young women from around the world.

The following are excerpts of their thoughts on how online discussions helped them grow their future endeavors. These women are from China, India, Nepal, Moldova, and Ukraine.

22.2.1   OFFERING EMPLOYMENT OPPORTUNITIES IN CHINA

Excerpt from Spring Zhou, Facebook Community Manager for a 3D printing company from Shenzhen, China

I met Joel Leonard in 2019 online via a facebook maker community. I was impressed by his passion to help people get jobs and I later invited him to speak on 3D printing for my company's online discussion and I was inspired to participate in his Bury 2020 International Time Capsule project. His project inspired me to want to pass on information so future generations can avoid some of the challenges we experienced during the pandemic.

After knowing him for about half a year, I just happened to share that I was ready to move on to work somewhere else and wanted to ask him for advice in my career. To Joel that was a challenge because he never been to mainland China and he wanted to help me.

He started sending me lists of local companies in ShenZhen and even helped me get two interviews with desired companies. I asked him, "How did he know so much about Shenzhen? I live here and never heard of many of these companies listed before." His response was priceless. Joel said "Nowadays with the internet and Google you don't need to know the answers, just the questions." I was able to attain one of those jobs and have been working here for almost 2 years. To me, Joel is more than a close friend, actually a caring family member to me now.

I hope more will look beyond their countries boundaries and help others around the world advance their careers and empower more women in STEM to our worldwide performance levels.

Joel Leonard's vision – Helping people from all over the world is such an honor and pleasure. When Spring expressed interest in getting employment, I thought, "What the heck, why not see if I can find her something?" That's when my journey began and even though I was not familiar with Shenzhen, I used my skill in research to find appropriately suited jobs for Spring. It was my intention to be able to provide some source of hope and a sense of empowerment to her.

Sometimes, critical skills such as knowing what to search for or how to search for it are not taught in schools. These are skills we pick up along the way and use them to navigate through life. By passing these on to Spring, she can now utilize them in her future and possibly share them with others who may be in similar situations. When we give wholeheartedly, we can share so much more with the world and help to empower others through our efforts.

22.2.2 DELIVERING CONTENT THROUGH THE MAKER FAIRES IN INDIA

Excerpt from Shweta Thapa from India

(now in the United Kingdom)

Believe it or not, I met Joel Leonard at the Milwaukee Maker Faire in 2019. Since we both have a love for 3D printing and the maker movement, we instantly struck a rapport and attended Maker Faires in Atlanta, Miami, Orlando. Joel helped me set up an office out of the Maker Depot Academy in Totowa, New Jersey. Joel was a strong advocate for helping women like myself uncover opportunities, and worked tirelessly to remove whatever obstacle was in our path.

I moved back to India, where I had Joel lead the online "Making It" conference and he got all of the attendees excited about the maker movement. While in India, I helped train girls in villages how to 3D print and am now pursuing my PhD in the United Kingdom.

I am looking forward to meeting Joel at future online conferences and Maker Faires. I wish more men would be as passionate about empowering women in STEM as Joel is.

Joel Leonard's Vision – It was a delight to help Shweta get connected in New Jersey and support her dreams to help more women enjoy the pleasures of designing, building, and creating their own 3D prints. Based on my extensive knowledge of setting up a business and being able to navigate the obstacles which come along with such, I was able to help her with her dream. Now, she gets to share that dream with everyone she comes into contact with through her travels.

Shweta's passion to help recruit and develop more young women in STEM is exactly what we need more of and need to support! She is helping to bring awareness to an industry which has not been heavily developed in the areas she visits. Through her sessions, she is opening the minds of young girls to teach them a skill they will be able to apply later on in life. I am so proud of her and the way she is inspiring the future generations.

22.2.3 PROVIDING NETWORKING OPPORTUNITIES IN NEPAL

Excerpt from Shikshya Gautam from Nepal

Over the last several years, I got to know Joel Leonard via Covid Weekly zoom calls to help makerspace survive the pandemic. I participated in the Bury 2020 International Time Capsule development event and have even gotten to visit in person with Joel Leonard during the Informa Markets Engineering (IME), Manufacturing Conference this past June. During this time, I had the opportunity to meet many women in STEM, shared my experience and learned with them which was a fruitful event. I got to meet a female engineer from Tesla and it was so inspirational meeting her.

Joel Leonard's Vision – Being able to connect with Shikshya and help her to become more aware of all the opportunities that await her was an absolute pleasure. She has taken her curious energy and channeled it into all of her projects. Her approach to new ideas and challenging the status quo will form the basis of her foundations as she strikes out on her own in this world.

22.2.4 GROWING TALENT IN NEPAL

Excerpt from Anu Thapaliya from Nepal

Shikshya Gautam, my friend from Nepal, told me about Joel Leonard from North Carolina. She and Joel encouraged me to speak at the IME Conference on how adding women to the workforce can help solve the ongoing shortage and maintenance crisis underway.

I shared my challenges in getting my engineering degree and the lack of women pursuing these roles. I hope that more will work like Joel to grow more talent around the world.

Joel Leonard's Vision – Being able to meet Shikshya and Anu was quite special as I never met anyone from Nepal before and both are extremely smart and dedicated to their educational journey. During the IME conference, I knew Anu was nervous and shy, and I was worried that she would not say much on the panel. Once she got over her initial anxiety, Anu got a burst of energy and started talking so much and offering some wonderful perspectives. She is a very sharp, passionate young professional who has so much to offer.

22.2.5 EXPANDING THE MAKER MOVEMENT TO MOLDOVA

Excerpt from Mariana Coscodon from Moldova

Youth Maker Club Moldova is a social project addressed to young people in technical vocational education. In the context of this project we organized a series of educational activities, training, practical workshops in order to increase employment opportunities for young people in the labor market and to promote maker movement across the country.

Having less experience in the field of maker movement, we started looking for models outside our country. We saw Joel Leonard's posts online about supporting the maker movement and we wanted to learn more. We have been talking to him for the last couple of years and even had him present his ideas at our own Maker Conference.

We believe that it is absolutely important to empower more women in STEM by connecting more with more online resources and mentorship provided by people like Joel.

Now more than ever, representatives of the maker community and supporters of STEM education need role models for inspiration. We are grateful for the input and expertise of Joel and the entire international community.

Joel Leonard's Vision – Not only is Moldova dealing with normal economic struggles every country faces but being a border country to Ukraine, they are also dealing with humanitarian crises. They are using drones to deliver medicine, food and supplies in remote and dangerous areas. There are groups called Techfugees who are using technology to locate support and provide supplies to survive this horrific conflict.

Mariana and folks in Moldova have huge hearts and want to help so many survive and grow their STEM skills, so that more women can create a better future beyond the war. The skills they learn now will transcend to everything they apply their minds to in the future. During these times, the inspiration provided by Mariana and her team will be the beacon of hope for the future.

22.2.6 MEETING MY MAKER FAMILY IN UKRAINE

Excerpt from Maryana Maslovska from Ukraine

(now in Chicago)

I came to the US and discovered Maker Depot Academy where I found more than tools and equipment but met my Maker Family. These makers guided and trained me to do more, create more and even produce PPE (Personal Protective Equipment) during covid to help more live during the Pandemic.

During the 2018 World Maker Faires in New York City, we met Joel Leonard whose booth was nearby. Joel was very impressed with the people and culture of this makerspace. Later Joel helped Maker Depot Academy get more community support by getting media coverage. Joel got us featured on PBS news and in numerous newspaper stories.

Sometimes it's hard to begin even when you have a lot of good ideas. You know how to do it theoretically but you have no practical knowledge. In Maker Depot, I met people who helped and guided me so I could create some beautiful things. These people became my Maker Family.

Joel Leonard's vision – What Maryana left out above is that she used to take the bus system and 3 different stops to get to the Maker Depot Academy. But Maryana made the time and effort to be there frequently because of the supportive and caring culture in that space which was created for its members.

Community Makerspaces around the country are wonderful resources for women to be empowered in STEM and grow their skills. Maryana made some wonderful items via laser engraver, electronics bench and 3D printer. She made extra money by selling custom jewelry and made Christmas gifts. We need to help other women find their maker family too. If you want to see a local makerspace – Google makerspace near me and arrange a visit, perhaps a makerspace will change your life too.

22.2.7 ENGAGING MORE WOMEN IN STEAM

Excerpt from Marita Garrett, Former Mayor of Wilkinsburg, PA, President and CEO, Civically, Inc.

I had the pleasure of meeting Joel in 2019 at the Nation of Makers Conference and immediately, when I heard him speak – I knew he was a fierce advocate for the maker community, especially engaging more diversity in the STEAM field.

We know that representation matters. When we don't see ourselves in these fields, it lends to a perception that we do not belong in these industries, which is the furthest from the truth. In order for more women, especially Black women, to seek careers in STEM – barriers to access must be removed in order to increase representation. Also, there must be mentorship and support in place to combat any biases faced by women.

Increase in gender equity is more than the right thing to do. It ensures there's more than one perspective and strengthens the culture of the STEM field.

Joel Leonard's vision – With all of the conflicts underway around the world, it is a very satisfying feeling to know you are helping advance other people in other cultures around the world. I hope that others emulate this practice so more women are empowered in STEM.

22.3 HOW DO WE HELP STUDENTS/EDUCATORS? – MAKESBOROUSA ACTIVITIES CASE STUDY

MakesboroUSA has been created to help remove the fog of the future from our youth and help them better understand viable career pathways in the resurging manufacturing sector. MakesboroUSA provides a manufacturing field trip experience on site, whether it's in the classroom, at a park, at an event, or anywhere you gather. Middle and high school students in rural and underserved communities are the core focus as these students need more guidance and options.

Industrial technology and programs have been created to offer fun, informative experiences for the youth involved. These currently consist of the Thorminator Challenge – assigning teams to build protections for eggs and a team egg smashers using a 50-pound hammer. We call these teams the Civil Engineers as they build targets. A team of Mechanical Engineers use the weapon, the Thorminator, and get to smash the egg protections. This activity helps give the students project-based learning experiences, opportunities to work together as teams, and during this challenge the students are informed of future career pathways coming to their area and what they need to do to get prepared.

After the eggs' protections are tested via the Thorminator smash, a team of Forensic Engineers get to evaluate the mess and see if the eggs survived the challenge. In the Acoustic Ultrasound Leak Detection demo and transmitter hunt, students learn how industrial sensors can help identify energy leaks and opportunities for companies to improve their energy efficiencies by plugging compressed air leaks.

Below are some excerpts from teachers. One of my favorite backhanded compliments from a group of teachers was, "Thanks for making us look bad!"

9/28/2022

To Whom It May Concern,

I would just like to take the time to describe a recent morning in my 8th and 7th grade Technology classes at Ledford Middle School. Mr. Joel Leonard of Makesboro USA visited our classes and guided my students on a technological adventure that … in my opinion … ALL students should have the opportunity to experience. Through several hands-on, project based learning challenges, the students "accidentally" gained some tech knowledge and career information while having a blast. Even this seasoned teacher had some fun! The next day, the majority of my tech students eagerly asked if and when Mr. Leonard was coming back. Unfortunately, I could not give them a solid answer. Wouldn't it be great to have students experience this type of learning and display this type of attitude much more often within their school lives? I think the answer is obvious. In closing, please help us educators make this happen on a more frequent basis. We need to find the funding that will allow Mr. Leonard to work his "magic" in as many schools and classrooms as absolutely possible. These experiences can do so much in preparing our children to face the working world that lies ahead of them.

Sincerely,
Mr. Jeff Teague
C.T.E. Teacher
Ledford Middle School

10/13/2022

To Whom It May Concern,

Joel Leonard from Makesboro USA came to visit my class today to get the kids excited and motivated to learn about CTE. He discussed Mechanical, Civil, and Forensic Engineering, then he involved the students in a hands-on activity related to these careers. He assigned jobs as the students walked in the classroom. He explained that they had the materials on their desks to protect an egg from being smashed by a giant hammer. Building the protective devices was an example of Civil Engineering. Using the hammer was an example of Mechanical Engineering. He had students check to see if the egg was smashed or not, which was an example of Forensic Engineering.

He had a discussion about jobs coming to NC and how they should take certain classes because there are programs that allow them to be apprentices, which could offer wonderful job opportunities. He also introduced them to and let them use an infrared thermal gun and a long range acoustic device.

The kids loved it! They were engaged, motivated, and kept asking when he was going to come back. He was engaged with the kids the whole time. I truly enjoyed seeing them so excited. I think it would be great if they could have more of these experiences.

Sincerely,
Mrs. Amanda Coppley
C.T.E. Teacher, North Davidson Middle School

22.4   WORKING TOGETHER TO EMPOWER AND GROW MORE TALENT

The society that scorns excellence in plumbing because plumbing is a humble activity and tolerates shoddiness in philosophy because it is an exalted activity and will have neither good plumbing nor good philosophy. Neither its pipes nor its theories will hold water.

— **John Gardner**

Parents used to advise their kids to study hard so they wouldn't have to work in a factory. Now they need to say study hard so you can get a big job in advanced manufacturing.

The first female reliability engineer I met said that women traditionally wanted to see their lives benefit humanity, like in medical and education professions. However, she saw a broader view and realized that manufacturing benefits humanity as well. She also loved that her research into reliability and ideas made businesses safer and products more affordable for consumers. To further empower women in STEM, we need to offer broader pathways and remove negative stigmas on manufacturing and gender biases by changing our cultural norms. We can encourage girls to weld, build robots, or even work with electrical/mechanical systems.

Women offer depth, detail, and sheer volume of workers which the industry desperately needs to continue to grow and replace retiring generations of existing workers. Families should encourage their daughters to tinker with cars, become entrepreneurial, and put less money into weddings and more money in funding their daughters' innovative business plans. They should encourage them to locate professional mentors from around the world, not just locally sourced. We need to boost and support their dreams to create a better tomorrow and pursue education with a new passion and desire to make the world a better place.

We need to engage more retirees to pass on their wealth of expertise to our youth by figuring out ways they can mentor via makerspace classes, or just periodic online chats. The important thing is that we cultivate professional interactions beyond generational groups and country boundaries where more people can learn the best concepts and skills to sustain and grow. This way, we can fill the thousands of unfilled technical jobs that our industries need filled to restore supply chains and address aging infrastructure challenges.

I hope that the examples detailed in this chapter help inspire more to mentor and be mentored so we all have a better, more prosperous tomorrow and all of us are empowered to do not just more with STEM but with our lives! Thanks so much for your time and attention to attempt to share best practices so more people can live more impactful lives and rise out of poverty.

If you were inspired by and want more information, feel free to reach out to me to discuss further at my Linkedin profile, https://www.linkedin.com/in/joelleonard makersmaker/

Let's all work to empower, to grow more talent and fill critical jobs with our youth who just need some guidance and mentorship.

ABOUT THE AUTHOR

A former vice president of the Association of Facilities Engineers, **Joel Leonard** has spent the past 30 years identifying, explaining and helping solve the problem of the maintenance crisis. His creative strategies to build awareness have taken him around the globe four times, where he has addressed international conventions, taught certification classes and gathered information to help others prepare for the maintenance crisis. As a result of his efforts, Joel was appointed to the United States Council on Competitiveness, a national think tank whose purpose is to work with Capitol Hill and the White House to create policy that results in legislation. International corporate and governmental leaders continually seek his advice. He has been interviewed on National Public Radio and CNBC numerous times. Currently, Joel is a workforce development consultant both in the United States and internationally for MakesboroUSA. He was recently called a great technology ambassador for New Zealand by a Member of the NZ Parliament, as he is currently helping to create strategies to boost NZ manufacturing. Joel is also the past chair of the National Defense Manufacturing Workforce Committee for NDIA.

Joel has been a longtime supporter of breaking stigmas and stereotypes that inhibit women from entering the skill tech and engineering sector. He created and administers the over-3500-member Wild Women Welders Facebook Group and wrote a song encouraging women to pursue the maintenance profession, called "Find Me a Maintenance Woman." Joel is thrilled to offer insight on how to empower more women to take on STEAM-related professions.

Index

Pages in *italics* refer to figures.

Taylor & Francis eBooks

www.taylorfrancis.com

A single destination for eBooks from Taylor & Francis
with increased functionality and an improved user
experience to meet the needs of our customers.

90,000+ eBooks of award-winning academic content in
Humanities, Social Science, Science, Technology, Engineering,
and Medical written by a global network of editors and authors.

TAYLOR & FRANCIS EBOOKS OFFERS:

A streamlined
experience for
our library
customers

A single point
of discovery
for all of our
eBook content

Improved
search and
discovery of
content at both
book and
chapter level

REQUEST A FREE TRIAL
support@taylorfrancis.com

 Routledge
Taylor & Francis Group

 CRC Press
Taylor & Francis Group

Printed in the United States
by Baker & Taylor Publisher Services